Earth and Environmental Sciences Library

Earth and Environmental Sciences Library (EESL) is a multidisciplinary book series focusing on innovative approaches and solid reviews to strengthen the role of the Earth and Environmental Sciences communities, while also providing sound guidance for stakeholders, decision-makers, policymakers, international organizations, and NGOs.

Topics of interest include oceanography, the marine environment, atmospheric sciences, hydrology and soil sciences, geophysics and geology, agriculture, environmental pollution, remote sensing, climate change, water resources, and natural resources management. In pursuit of these topics, the Earth Sciences and Environmental Sciences communities are invited to share their knowledge and expertise in the form of edited books, monographs, and conference proceedings.

Gehan Mohamed El Zokm

Ecological Quality Status of Marine Environment

Metal-Sulfide Models; Significance, Mobility, Mechanisms and Impacts

Springer

Gehan Mohamed El Zokm
Marine Chemistry Lab, Marine
Environment Division
National Institute of Oceanography
and Fisheries
Cairo, Egypt

ISSN 2730-6674 ISSN 2730-6682 (electronic)
Earth and Environmental Sciences Library
ISBN 978-3-031-29205-7 ISBN 978-3-031-29203-3 (eBook)
https://doi.org/10.1007/978-3-031-29203-3

This Springer imprint is published by the registered company Springer Nature Switzerland AG
The registered company address is: Gewerbestrasse 11, 6330 Cham, Switzerland

Preface

Techniques to evaluate metal bioavailability in sediments have received more consideration as a way to assess sediment toxicity because of the potential environmental effects. One of the most reactive and significant chemical anions is acid volatile sulfide (AVS), which regulates the toxicity and ecological concerns associated with heavy metal impacts in aquatic environments. However, copper, cadmium, lead, zinc, and nickel cationic forms have strong relationships between their bioavailability and AVS level. The worldwide geographic variability of AVS for aquatic environment is still unidentified. Heavy metals in interstitial water can be reduced or removed by sediments that (co)precipitate to Mn and Fe in MnS or FeS minerals. To evaluate the metals (SEM) that simultaneously extracted with AVS extraction, a special technique is employed. It's made under moderately to highly reducing circumstances. AVS is also a critical link in understanding the dynamic of biogeochemical cycles. Acid volatile sulfide models have recently received more attention as effective approaches for evaluating sediment quality. The equilibrium partitioning theory (EqP) and the chemistry of metal–sulfide interactions are used in SEM-AVS models to predict toxicity. AVS models have been documented as one of the most important techniques for forecasting the state of the aquatic environment. This book aimed to provide a critical deep insight into various AVS-SEM models as predictors for assessing the ecological impact of heavy metals to aquatic environment, including the theories behind these models, descriptive equations, their mode of action, methodology, efficiency, applicability, and statistical analysis, as well as distinctions with other pollution assessment techniques. AVS/SEM studies in Egypt are compared to the aquatic environments around the world. Different case studies are discussed. The challenge with AVS-SEM models comes from the AVS components that are dramatically sensitive to oxidation conditions. From a future perspective, management is advised to overcome the difficulties within that technique such as accurate handling procedures and different approaches of sampling; onshore and offshore. Sequential leaching strategies, especially geochemical analysis, and knowledge of the interactions and significance of AVS in the marine sector, especially toxicity tests (bioassays), are recommended.

This book is directed to (1) delineate a conceptual framework as a deep and comprehensive view of one of the most important mechanisms used in the assessment of the aquatic environment, which is the evaluation of heavy metal pollution using AVS–SEM technique, (2) monitor the technology's description, including the presentation of terms and definitions used in these models, (3) highlight the sulfur cycle and factors controlling the presence of AVS with special reference to the equations governing the distribution of sulfur species, (4) clarify oxidation-reduction chemistry's role in AVS systems, (5) to present different methodologies used in evaluating this technique, (6) provide a critical understanding of various AVS-SEM models used in assessing the status of the aquatic environment, including the theories underlying these models, their mechanisms, efficiency, applicability, and statistical analysis to evaluate AVS models, as well as their comparison with other pollution assessment techniques, (7) use AVS as a tool to classify the lakes (healthy area-eutrophic area—warning area—critical area), (8) give an overview of the studies that used this technique, especially those that were performed in the Egyptian aquatic environment and their outputs, and (9) show the update and limitations of that technique and suggest some proposals to overcome those difficulties in the future.

Last but not least, I want to thank all Springer team, in general, for their kind cooperation and in particular, Prof. Abdelazim Negm, the editor of the EESL book series for his generous support and review of the manuscript.

Cairo, Egypt
2023

Gehan Mohamed El Zokm

Contents

About the Author

Dr. Gehan Mohamed El Zokm is a professor at the Marine Chemistry Department and the leader of the halogen, sulfide modeling, and surfactant groups at the National Institute of Oceanography and Fisheries, Mediterranean Sea Branch in Egypt. She earned her B.Sc., M.Sc., and Ph.D. in the Chemistry Department at the Faculty of Science of Alexandria University in Alexandria, both of which concerned heavy metal chemistry (focusing on preconcentration and fractionation tools). She is a professor of environmental chemistry in the marine chemistry laboratory at the Marine Environment Division. Her research interests are in analytical and instrumental chemistry. Statistical analyses were the most prominent features of her research, which was conducted using the most recent software packages especially SPSS. Her papers discuss the topics of eutrophication, pollution as well as the responses of marine biota to these two phenomena. Her study focuses on the mobility and fluxes of various parameters within marine environment components. Extraction of natural products with many pharmaceutical applications and nutritional studies are her recent projects. Her research is also focused on water treatment and monitoring of pollutants (heavy metals- phenol-hydrocarbons and pesticides) as well as sulfide models dynamic. She has 52 research papers and two books with an H-index of 16. Factorial design is applied for water treatment experiments. She reviews a lot of papers for many important publishers (Elsevier, Springer, and Taylor & Francis). She supervised and headed several projects and postgraduate students, participated in many conferences and workshops, and has several activities.

Abbreviations

AAS	Atomic absorption spectroscopy
AF	*At. ferrooxidans* from iron (II)-oxidizing bacteria
AMD	Acid mine drainage
AMP	Adenosine monophosphate
ANME	Anaerobic methane oxidizing
APS	Adenosine 5′-phosphosulfate
ARD	Acid rock drainage
ATP	Sulfate adenyltransferase enzyme (sulfurylase enzyme)
AVS	Acid volatile sulfide
AVS-SEM	Models described acid volatile sulfide and simultaneously extracted metals
BCR	European Community Bureau of Reference sequential extraction procedure for metals in sediment
BCR	The modified European Community Bureau of Reference (Three-step sequential extraction procedures)
BIO	Bioaccumulation
BLM	Biotic ligand model
BOD	Biological oxygen demand
C and H	Cold (C)/Hot (H) HCl
Cd	Cadmium
CH_2O	Organic matter
Co	Cobalt
COD	Chemical oxygen demand
Corg	Organic matter
C_{org}	Organic matter concentration

CRS	Pyrite sulfur (CRS; The chromium (II)-reducible sulfur)
C_{SQC}	Sediment quality concentration
Cu	Copper
C_{WQC}	Water quality concentration
Cysteine	Sulfur-containing amino acid
DIET	Direct interspecies electron transport
DMS	Dimethyl sulfide (CH_3SCH_3)
DMSP	Dimethyl sulfonio propionate

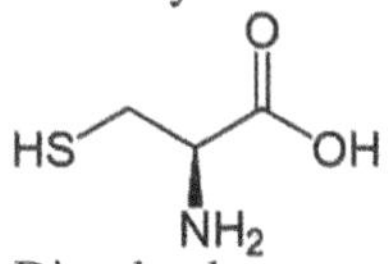

DO	Dissolved oxygen
Eh	Oxidation-reduction potential (mV: milli volt)
Enzymes	(1) APS reductase, (2) PAPS reductase, (3) sulfite reductase
EqP	Equilibrium partitioning model
ERL	Effect range low
ERM	Effect range median (USA)
ES	Elemental sulfur
ESBs	Equilibrium partitioning sediment benchmarks
Fe	Iron
Fe/MnOx	Iron and manganese oxides
Fe^{2+}	Ferrous ion
Fe_3S_4	Greigite
FeS_2	Pyrite
FeS_{aq}	Aqueous ferrous sulfide
fN2	Flow rate of the extracted N2 in purge and trap method in AVS extraction
f_{oc}	Fraction organic carbon in the sediment
FOM	First-order model
H_2S	Hydrogen sulfide
HS^-	Hydrosulfide
ICP-AES	Inductively coupled plasma-atomic emission spectrometers
ISE	Ion-selective sulfide electrode
ISQG	Interim sediment quality guidelines
kJ/mol	Kilo joule/mole

K_{OC}	Organic carbon–water partition coefficient
Kp (L/kg)	The partition coefficient between sediment and interstitial water
Ksp	Solubility product
LC_{50}	The concentration causing 50% death
Lf	*L. ferrooxidans* from iron(II)-oxidizing bacteria
LFe	Labile Fe
MB	Methylene blue
MC	Moisture content
MDR	Mixed-diamine reagent for trapping sulfide
Me^{2+}	Cationic metal
MeS	Metal sulfide
MLR	Multiple linear regression
Mn	Manganese
Ni	Nickel
Org-S	Organic sulfur
OSM	Organo-sulfur molecules
PAP	Adenosine 3′-phosphate 5′-phosphate
PAPS	Adenosine 3′-phosphate 5′-phosphosulfate
PEL	Probable effect level (Australia and New Zealand)
PELq	Mean PEL quotient
PPi	Pyrophosphate enzyme
PSD	Particle size distribution
PTE	Potentially hazardous element
R1	Reactor no. 1 in purge and trap method in AVS extraction
R2	Reactor no. 2 in purge and trap method in AVS extraction
RIS	Reduced inorganic sulfur
ROS	Degree of sulfidation
RQ	Risk quotient
RQPEL	Risk quotient based on total metal and PEL
RQSEM-PEL	Risk quotient based on SEM and PEL
RQSEM-TEL	Risk quotient based on SEM and TEL

RQTEL	Risk quotient based on total metal and TEL
S^0	Elemental sulfur
S^{2-}	Sulfide
$S_2O_3{}^{2-}$	Thiosulfate
SAOB	Sulfide antioxidant buffer
SEM	Simultaneously extracted metals
SEM_{Zn}, SEM_{Cu}, SEM_{Pb}, SEM_{Ni} and SEM_{Cd}	Simultaneously extracted zinc, copper, lead, nickel, and cadmium
$SO_3{}^{2-}$	Sulfite
$SO_4{}^{2-}$	Sulfate
SPM	Shrinking particle model
SQC	Sediment quality criterion
SQGVs	Sediment quality guideline values
SRB	Sulfate-reducing bacteria
SRP	Sulfate-reducing prokaryrotes
SWI	Sediment water interface
TEL	Threshold effect level
TEL	Threshold effect level and PEL: probable effect level (Chinese)
TOC	Total organic carbon
TRS	Total inorganic reduced sulfur
Zn	Zinc
$[M]_R$	Residual metal fraction by sequential extraction technique
$\Delta G°$	Free energy of the reaction
ρ_S and ρ_{IW}	Concentration of metal in the sediment (ρ_S) and interstitial water (ρ_{IW})

List of Figures

List of Tables

Chapter 1
Introduction to the Significant Impact of AVS on Controlling the Metal Toxicity Regarding Sulfur Cycle

Heavy metals have been categorized as a serious and hazardous anthropogenic contaminant in marine environments [1–8, 10, 11; 77, 78]. Heavy metal contamination in aquatic systems is a worldwide challenge that has received huge attention recently. Heavy metals are non -biodegradable, poisonous, and accessible to store in organisms in the environment. Toxic heavy metals are primarily delivered into the aquatic ecosystem through various natural and human sources. Weathering processes (rocks and soils) and particle deposition in the atmosphere are the primary natural sources. Due to human activities such as industrial waste, farm discharges, mining operations, and sewage runoff, heavy metal levels in aquatic environments have risen sharply [1, 9–16, 76]. Keeping an eye on heavy metal pollution along the marine coast has been confirmed by several studies performed all over the world. The coastal sediments are one of the most significant sources and sinks for these contaminants, which can have a negative impact on the ecosystem and global health [2, 4; 77, 78]. Investigating the harmful effects and biogeochemical pathways of the metal requires knowledge of how the metals exist. The chemical formula of metals regulates their mobility and ecotoxicity [6]. Recently, approaches and models for prioritizing tasks, creating standards for environmental quality, and assessing risks have been established [3]. The reactive sulfide that can be volatilized by cold HCl extraction at a concentration of 1 or 6 M is known operationally as AVS. It has been considered as an indicator of toxicity in both freshwater and marine sediment due to its potential to bind to heavy metals and settle them in aquatic environmental sediments. AVS is a biogeochemical reactive fraction sulfide in sediment with a significant affinity for cationic metals (Pb, Zn, Cd, Ni, Cu). Sediments act as a secondary pollution source when heavy metals that have deposited in the sediment phase might be released back into the waterbody and the atmosphere when the environment changes [5]. Iron and sulfur along with oxygen and carbon are four elements that dominate diagenetic processes and the redox states of the earth's surface. As a result, their cycling is crucial to a number of signifi(cant processes occurring all over the world [17]. Numerous types of sulfur, including sulphate (SO_4^{2-}), reduced inorganic sulfur (RIS), and organic sulfur (Org-S), are found in nature. Acid volatile sulfur (AVS), pyrite sulfur (CRS), and elemental

G. M. El Zokm, *Ecological Quality Status of Marine Environment*, Earth and Environmental Sciences Library, https://doi.org/10.1007/978-3-031-29203-3_1

sulfur (ES) are the major components of RIS. When viewed from the perspective of the extraction process, Org-S includes chromium reducible organic sulfur, chromium non-reducible organic sulfur, and humus sulfur [18]. Norman et al. [19] classified Org-S as carbon-bonded sulfur (C–S) and ester sulfate (C–O–S). Numerous biotic and abiotic mechanisms regulate the transformation of sulfur species in sediments.

1.1 Definition of AVS and SEM

Technically, AVS stands for a fraction containing comparatively labile Fe and Mn mono sulfides as the predominant components. Zn, Cd, Ni, Cu, and Pb displace Fe and Mn in their sulfide minerals, and the new metal sulfides are less bio-available to aquatic organisms [20–23]. According to reports, AVS in sediment is significantly correlated with the bioavailability of several divalent metals. $\sum$SEM are the metals extracted during AVS extraction technique. The chemical concept for this reaction is that divalent metals displaced iron in FeS rapidly to build more MeS (Eq. 1.1)

$$Me^{2+} + FeS(s) \rightarrow MeS(s) + Fe^{2+} \tag{1.1}$$

Metal toxicity in sediments was eliminated when AVS exceeded the total cationic ions since in molar concept. This assumption is based on the equilibrium partitioning theory used to study the interaction between AVS and metals. The AVS—metal linkage produces thermodynamically stable metal sulfide precipitates lowering metal availability and mobility [24]. Metals can flux between the dissolving form and sediment elements such oxyhydroxides or organic detritus when SEM exceeds AVS (excess metals). Metal concentrations in interstitial waters depend on $\sum$SEM, the degree of solubilities of MeS; Pb, Ni, Cd, Cu Zn, and metal flux with sediment constituents rather than AVS as organic matter and oxides (Fe/MnOx) [22].

1.2 Impacts of Sulfide on Metal Toxicity

According to [25], AVS is a meta-stable fraction made up of a combination of dissolved (S^{2-}, Fe^{2+}, and polysulfide) and solid sulfur species (mackinawite and greigite). In most cases, AVS is discovered in sediments with a redox potential of less than 100 mV. Sulfate-reducing bacteria, on the other hand, aid in the production of AVS in the marine environment [26–28].

Significant sulfate-reducing bacterial activity is promoted by anoxic conditions, sulfate supply, and labile organic matter [29–31]. Metal bioavailability and toxicity in aquatic environment are influenced by the following factors: (a) binding phases; AVS, particulate organic carbon, iron, and manganese oxyhydroxides and (b) physico-chemical properties of the bottom and pore water, such as salinity, pH, redox potential (Eh), hardness, and ligand availability. and (c) the degree of sulfide sensitivity and attitudes of benthos, including way of life and previous exposure background [32–34].

When evaluating heavy metal threats in sediments, three distinct possible processes occur concurrently but not inevitably sequentially: dissolution, conversion to bioavailable forms, and turnaround to non-bioavailable forms [35]. According [36], Fe/Mn-oxide, clay minerals, sulfides and organic compounds, are the main forms to precipitate and absorb metals.

1.3 AVS as a Predicter for Metal's Future Bioavailability and Toxicity

AVS and SEM criteria can be used to predict a metal's future bioavailability. The apportionment of AVS concentrations varies spatially and temporally, and the levels of AVS increase with sediment depth until a certain depth [34, 37–39]. The low AVS values in surficial sediment are due to bio-irrigation exposure and oxygen penetration from surface waters, which cause aerobic oxidation of the sulfide and increase the metal mobility capacity of sediments [22]. Reactive AVS and Fe were replaced by crystalline components, such as pyrite, inside deeper sediments, resulting in lower amounts of AVS and Fe [34]. The relevance of AVS as a factor influencing ambient variables that influence metal behaviour has been shown in anoxic sediments [1, 40–42]. As sulfide phases in anoxic sediment have poor solubility, AVS concentrations are expected to be high enough to react with cationic metals and reduce their long-term impacts [22].

1.4 AVS Sources; Aller's Relationships for Natural Sulfidic Systems

With the help of Aller's research [43], the kinetics of the FeS system has been better understood. It has been evidenced that oxidants play a crucial part in controlling AVS to pyrite ratios. The simplest mathematical representation of the for the chemical reactions affecting the concentration of FeS is as follows:

$$C_{FeSt} - C_{FeS0} + \left(R_{formation} + R_{oxidation} \mid R_{pyritizationt}\right) \quad (1.2)$$

where R is the net rate of production or consumption, C_{FeSt} is the sum of the concentrations of FeS_s and FeS_{aq}, and C_{FeS0} is the concentration of FeS at time (t) = 0. FeS forms at an incredibly rapid rate, with typical times in the millisecond range (Rickard 1997). S(–II) generation by bacteria occurs at a slower rate than FeS formation. This indicates that FeS is generated at the same rate as the availability of S(–II).

Fig. 1.1 A simple "black box" model for AVS [25]

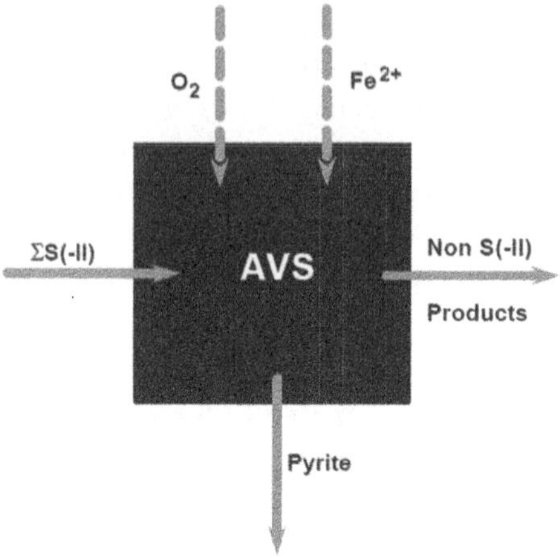

1.5 Sulfate Reducing Prokaryrotes (SRP) Behaviour

According to [44], the SRP consists of about 18 genera that utilize sulfate during anaerobic respiration as a final electron acceptor. The production of APS enzyme from ATP and sulfate, is a crucial step in sulfate reduction by SRP to give $SO_3(-II)$ which can be reduced more easily, with the assistance of APS to S(–II) [25].

- **Degree of sulfidation (ROS)**
- According to Morse [45], The majority of total inorganic reduced sulfur (TRS) typically adds to the S reservoir is pyrite-S.

$$\text{DOS} = \frac{\text{TRS} + \text{AVS}}{\text{TRS} - \text{AVS} + 2\ \text{Leachable} - \text{Fe}} \tag{1.3}$$

AVS = AVS-Fe

Reactive-Fe = the Leachable-Fe-AVS-Fe

Under steady-state settings, there won't be any measurable difference in the AVS creation and loss to pyrite and oxidation will balance one other out. Therefore, rather than being dependent on the absolute rates themselves, the concentration of AVS depends on the previous degree of disequilibrium in these processes' rates and the length of this disequilibrium (Fig. 1.1).

1.6 Equilibrium of AVS System in Sediment

The balance of AVS production and loss in sediment was used to determine the equilibrium level of AVS in sediment. The rate of supply and reduction of SO_4^{2-} organic matter decomposition and redox potential control the AVS equilibrium mechanism. Sulfate-reducing bacteria react with organic molecules as described in Eq. 1.4, giving sulfide ions in the presence of sulfate ions.

$$2CH_2O + SO_4{}^{2-} \xrightarrow{(SRB)} 2CO_2 + S^{2-} + 2H_2O \tag{1.4}$$

When metals that are sensitive to redox potential (Fe, Mn, Zn, Cu, Ni, Cd, Co) react with sulfide, minerals and colloids are formed. The AVS technique is applied to measure the concentration of sulfide evolved as hydrogen sulfide gas and SEM after the addition of hydrochloric acid. The AVS technique produces a collection of dissolved sulfur species present in sediment pore water and weakly crystallized sulfides that are mainly mackinawite minerals (FeS) [25, 46, 47]. Sulfur is an important redox element involved in numerous biogeochemical processes [48]. Chemical properties, especially anoxic biological sediments from finely ground depositional zones, is being used to explain why there are higher concentrations of AVS in the aquatic environment. Seasonal temperatures have a significant impact on the breakdown of organic matter, which has an impact on the amounts of AVS in sediment [23]. AVS concentrations in lake sediments and lake water temperature are known to be correlated [1, 23]. Significant fluctuations in AVS concentrations in sediment have been discovered by researchers in lakes with a periodic basis anoxic hypolimnion [49]. Low-flow, low-gradient water bodies can have ambient concentrations of AVS due to hydrological phenomena including stream flow. In anoxic and semi anoxic sediments, sulfur forms in the sedimentary environment function are a main controller for trace metal mobility. Howarth [50], found that sulfur's redox cycle has an impact on sediment-bound metallic contaminants. Microorganisms can also oxidize sulfides to produce intermediate sulfur species such as thiosulfate, sulfite, and pyrite, as well as elemental sulfur and sulfate [51–53].

1.7 In-Depth Sulfur Cycle Analysis Regarding Different Literature

The biogeochemical transformation and recycling of sulfur, one of the basic nutrients, is necessary to maintain the functionality of aquatic ecosystems (Fig. 1.2) [54]. Sulfur biogeochemistry, has a direct impact on the restoration of organic matter, water acidification, nutrient cycle, and metal bioavailability in aquatic environments [53]. Only a few of the sulfur species found in aquatic sediments are sulfide, elemental sulfur, sulfate, and organic sulfur compounds. Microbial behaviuor, Eh, potential, organic carbon, and acidity all have an effect on these species' transformations [53]. Sulfate can be produced by microorganisms that break down sulfur [55]. Anoxic and anaerobic conditions enhance sulfate to sulfide reduction by SRB [56].

To understand the AVS's role in the environment, the sulfur cycle must be studied carefully. Much literature describes the sulfur cycle [52, 57–59]. According to [52, 54], the biogeochemical cycle of sulfur in aquatic sediments is extremely sensitive to the redox balance and the composition of the seas (Fig. 1.2). It's complicated since

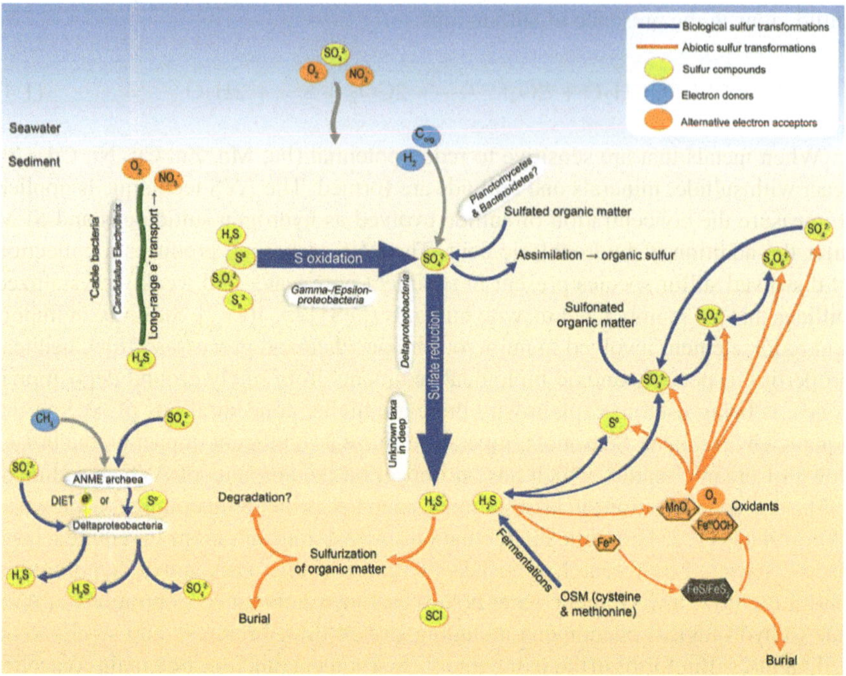

Fig. 1.2 Definitional portrayal of the sulfur cycle in sediments with marine origin, including important reactions of inorganic and organic sulfur species, and evolutions of intermediate oxidation state sulfur compounds (sulfur cycle intermediates, SCI). The physiologically driven sulfur changes represented by the blue lines can also be a part of disproportionation processes. Orange lines represent abiotic reactions. Yellow eclipses are a symbol for inorganic sulfur compounds. Additional electron acceptors are displayed in yellow ellipses, and additional electron donors are displayed in blue ellipses. OSM = organo-sulfur molecules, Corg = organic matter. DIET = direct-interspecies electron transport. ANME = anaerobic methane-oxidizing [54]

there are so many different oxidation states. Furthermore, both chemical and bacterial changes occur at considerable rates. Between the most reduced H_2S (− 2) and the most oxidized SO_4^{2-} (6), the sulfur cycle involves eight electron oxidation/reduction processes. In many bacterially mediated processes, it acts as an electron donor or acceptor. Sulfur exists in all oxidation states in the environment, although it is most commonly found as elemental (S^0), negatively divalent and positively hexavalent (sulfate). Each of these forms can only be utilized by certain species, which then change it into another form that can be used by other creatures [60, 61]. Iron sulfide (mackinawite, greigite, amorphous FeS, pyrrhotite, or troilite), manganese sulfide, pyrite (iron bisulfide), or organic sulfide make up the bulk of sedimental sulfide ions [62, 63].

Glöckner et al. [59], described the sulfur cycle (Fig. 1.3). A wide range of oxidation states (− 2 to + 6). Transitions from one redox state to the next are catalyzed by microorganisms and, in certain cases, chemical processes. The "mini-sulfur cycle" refers to the 2-electron transition between sulfide (− 2) and elemental sulfur (0) by an oxygenic phototrophic (photosynthesis) or colourless sulfur bacteria (chemosynthesis). It is subsequently reduced back to sulfide by anaerobic respiring bacteria. Sulfur is also oxidized to sulfate. Colorless sulfur bacteria may also oxidize sulfide straight to sulfate. In one of three strategies, sulfate can be absorbed for structural cell material, reduced to sulfide, or deposited as inorganic and organic minerals in sediments. Sulfide is re-mineralized from organic molecules that have died. During these sulfur conversions, several intermediate redox states emerge. Bacteria can oxidize SO_3^{2-} and $S_2O_3^{2-}$ intermediates to SO_4^{2-} and produce electrons that generate sulfide ions. Inorganic fermentation, or disproportionation, is another name for this process.

According to [58, 61], plants and certain microbes manufacture organic molecules from the sulfate. These are subsequently digested by mammals and used as a substrate by microorganisms. Hydrogen sulfide is produced and discharged into the environment during this breakdown. Colorless sulfur bacteria aerobically convert it to elemental sulfur or sulfate (thiobacilli and fibrous sulfur bacteria from the genera Thiothrix, Beggiatoa, etc.). Hydrogen sulfide is oxidized anaerobically by phototrophic sulfur bacteria, yielding elemental sulfur (Fig. 1.4). Other studies described the sulfur cycle in aquatic sediments (anthropogenic) [61, 64–66]. Human activities influence the warming of the ocean and nutrients in coastal waters impact the benthic marine sulfur cycle (Fig. 1.4). This increases photosynthetic productivity

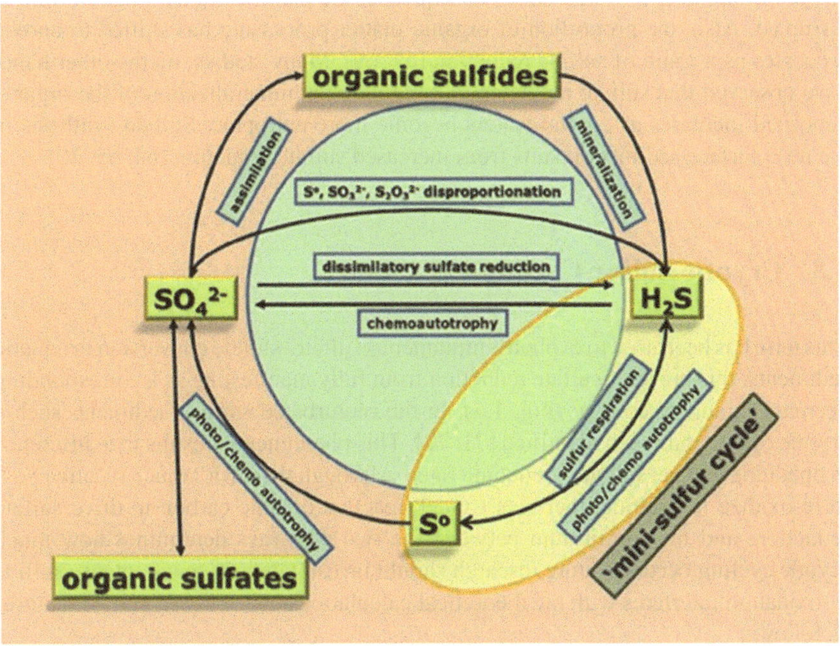

Fig. 1.3 Sulfur cycle according to [59] (doi: https://doi.org/10.13140/RG.2.1.5138.6400, http://archives.esf.org/fileadmin/Public_documents/Publications/MarineBoard_PP17_microcean.pdf)

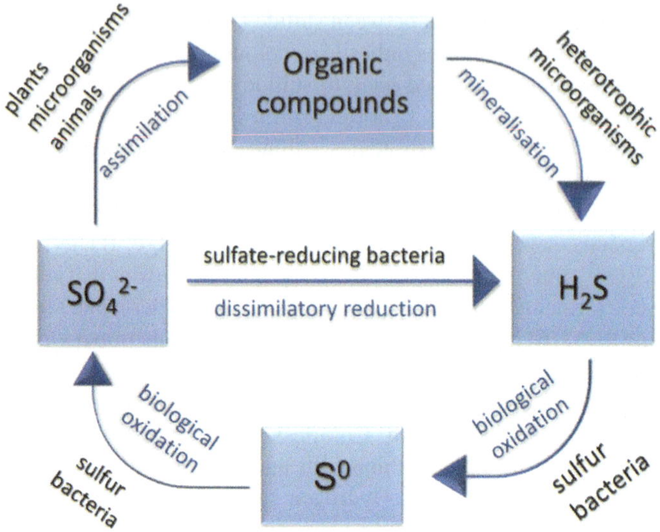

Fig. 1.4 Scheme of the microbial sulfur cycle [58, 61]

and organic matter export to the seabed, which is accompanied by a shortage of oxygen in the deeper water. As a consequence, the organic matter's equilibrium is disrupted. Also, the proportion of organic matter processing has shifted to anoxic processes as a result of sulfate reduction [67–69]. Many studies, on the other hand, have observed that sulfate reduction can regulate the mineralization of the organic deposited increases as coastal waters become more eutrophic. Sulfide synthesis in the near-surface sediment results from increased sulfate reduction [64, 65, 70].

1.8 Cryptic Sulfur Cycling

This term has been used to explain simultaneous sulfate–sulfide conversion in aquatic sediments, this prevents sulfate reduction from fully manifesting as a corresponding decrease in sulfate content (Fig. 1.5). In the bioturbated surface sediment, such a cryptic cycling has been identified [71, 72]. This phenomenon results in a difference in operating and net sulfate—reducing rates. Although there isn't much reactive Fe^{3+} to re-oxidize the sulfide, there isn't much reactive organic carbon to drive sulfate reduction, and the equilibrium between the two pathways determines how much cryptic cycling occurs. Future research should investigate how much intense sulfide re-oxidation interferes with the theoretical calculations concavity of sulfate features as follows:

$$17H_2S + 8FeOOH + 2CO^2 \rightarrow SO^{2-} + 8FeS_2 + 2CH_4 + 16H_2O + 2H^+ \quad (1.5)$$

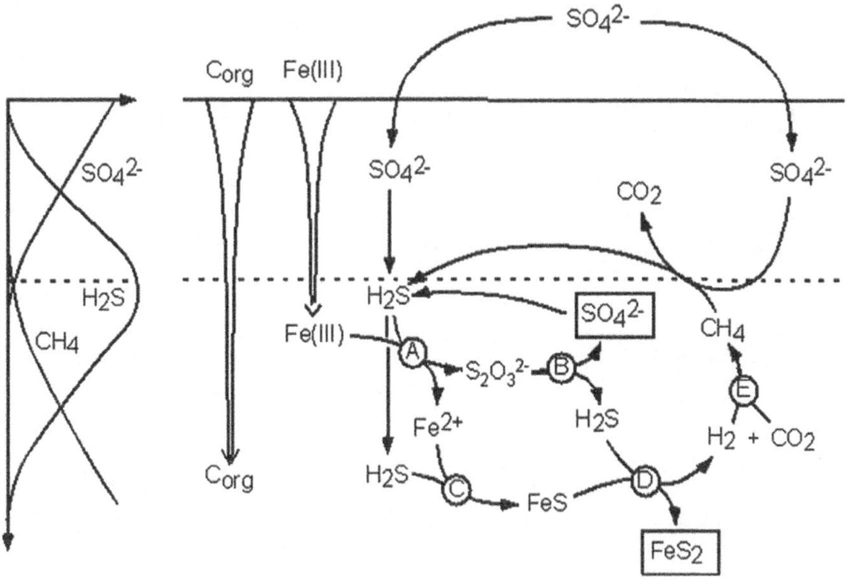

Fig. 1.5 The cryptic sulfur cycle driven by iron in the methane zone of marine sediment: a conceptual framework [72]

1.9 Sulfur Diagenesis in Marine Sediments (Assimilatory–Dissimilatory)

Sulfate is the starting material for two anaerobic processes known as assimilatory and dissimilatory sulfate reduction (Figs. 1.6, 1.7 and 1.8). In particular, microorganisms use these mechanisms to obtain energy for their activities. The ultimate product of the assimilatory sulfate reduction is cysteine, whereas sulfides is the terminal product of the dissimilatory sulfate reduction. The enzymes involved in the catalysis of the processes are another difference between assimilatory and dissimilatory sulfate reduction [73, 74].

Figure 1.6 depicts the cycling of sulfur in benthic sediments. Mandatory and fermentative biogeochemical cycling of sulfur aerobic organisms quickly consume all available oxygen. Then, transitioning to anaerobic processes that use, in that order, Fe/MnOx, ferric nitrates, and sulfates.

1.10 Effects of Climate Change on Marine Biogenic Sulfur Cycle

Dimethyl sulfide (DMS) is a greenhouse gas generated from a variety of organisms contribute for up to 80% of biogenic sulfur emissions worldwide. It is a vital

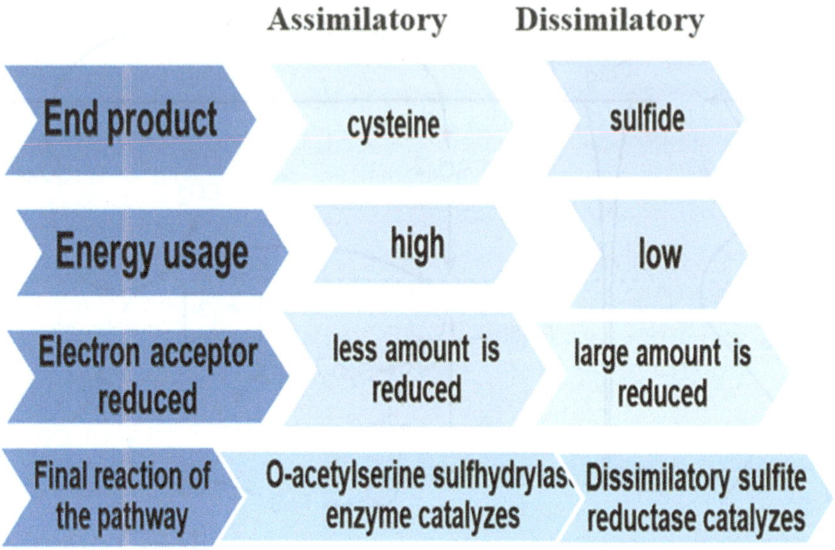

Fig. 1.6 An infographic on assimilatory and dissimilatory sulfate reduction differences (https://www.differencebetween.com/difference-between-assimilatory-and-dissimilatory-sulphate-reduction/)

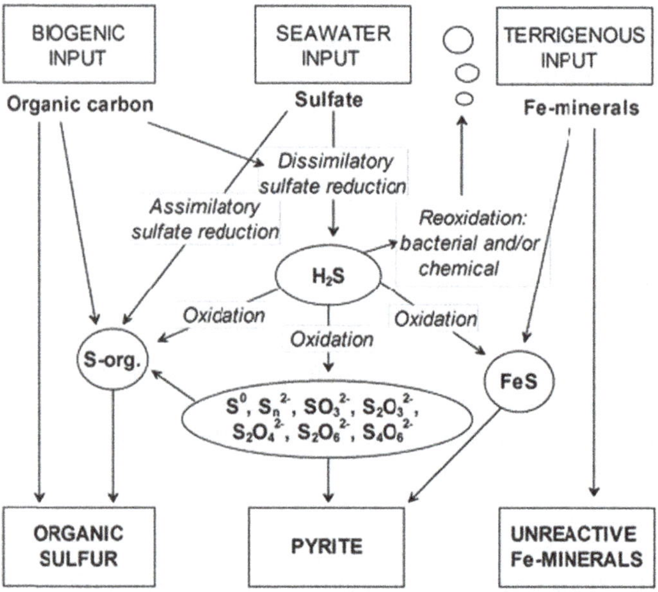

Fig. 1.7 Schematic diagram of the sulfur diagenesis in marine sediments [74]

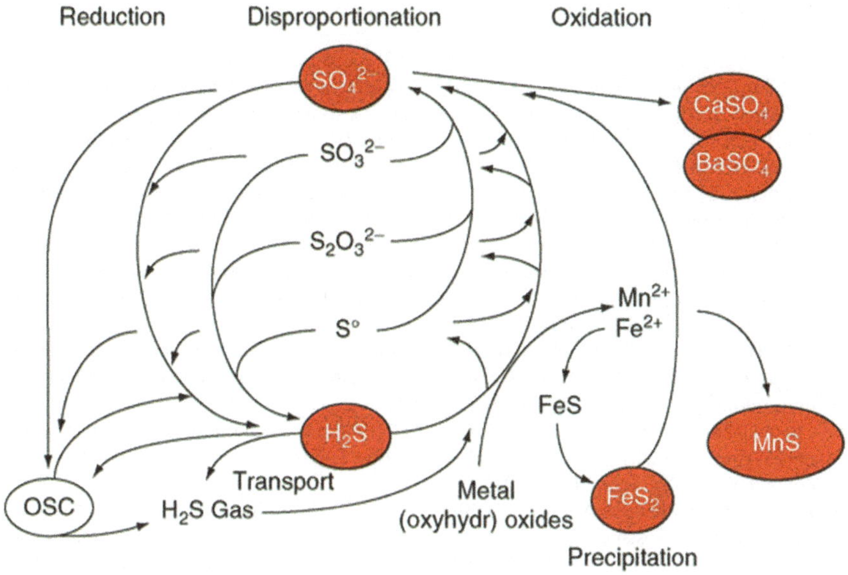

Fig. 1.8 Sedimentary sulfur cycle in a simplified form and its relevance to certain metals (neglecting, e.g., transformation of thionates, polysulfides, thiols, and methane). OSC: Organic sulfur compounds [73]

ingredient of the marine sulfur cycle. The DMS cycle begins with the non-gaseous precursor dimethyl sulfonio propionate (DMSP), which is entered to the water bodies by various processes in food web such as zooplankton grazing. Microbes use two known pathways to quickly convert this dissolved DMSP pool to produce DMS: demethylation (releasing methanethiol). A portion of the newly created DMS is evacuated into the atmosphere, where it quickly oxidizes and helps to create sulfate aerosols, which may have an impact on local climate change. DMS is the most prevalent type of oceanic volatile sulfur (S) and is a primary biological source of reduced sulfur (S) in the atmosphere. DMS has a critical role in supplying carbon and sulfur to marine environment. Due to its similarity to phosphorus in algal cellular stoichiometry, sulfur is a crucial component for marine primary production. Because they can absorb sulfate (SO_4^{2-}), which is present in large quantities in the ocean, marine phytoplankton are therefore essential components of the global S cycle. The capacity of phytoplankton to manufacture cellular DMSP enables the transfer of S from the sea to the air [75].

References

1. El Zokm GM, Okbah MA, Younis AM (2015) Assessment of heavy metals pollution using AVS-SEM and fractionation techniques in Edku Lagoon sediments, Mediterranean Sea. J Environ Sci Health Part A Toxic/Hazard Subst Environ Eng 50:571–584. https://doi.org/10.1080/10934529.2015.994945
2. El Zokm GM, Okbah MA, Kh E-S (2021) Integrated approaches to assess water quality in two spots along the western Mediterranean Sea, Egypt. Chem Ecol J 37(6):493–514. https://doi.org/10.1080/02757540.2021.1892657
3. Ahmed AY, Abdullah MP, Siddeeg SM (2022) Environmental hazard assessment of metals in marine sediments of Sabah and Sarawak, Malaysia. Int J Environ Sci Technol. https://doi.org/10.1007/s13762-022-04514-z
4. Nour HE, Helal S, Wahab MA (2022) Contamination and health risk assessment of heavy metals in beach sediments of Red Sea and Gulf of Aqaba, Egypt. Mar Pollut Bull 177:113517. https://doi.org/10.1016/j.marpolbul.2022.113517
5. Fan Y, Chen X, Chen Z, Zhou X, Lu X, Liu J (2022) Pollution characteristics and source analysis of heavy metals in surface sediments of Luoyuan Bay, Fujian. Environ Res 203:111911. https://doi.org/10.1016/j.envres.2021.111911
6. Zhang S, Ye H, Zhang A, Ma Y, Liu Q, Shu Q, Cao X (2022) Pollution characteristics, sources, and health risk assessment of heavy metals in the surface soil of Lushan Scenic Area, Jiangxi Province, China. Front Environ Sci 10:891092. https://doi.org/10.3389/fenvs.2022.891092
7. Xie S, Jiang W, Sun Y, Yu K, Feng C, Han Y, Xiao Y, Wei C (2023). Interannual variation and sources identification of heavy metals in seawater near shipping lanes: evidence from a coral record from the northern South China Sea. Sci Total Environ 854:158755. ISSN: 0048-9697. https://doi.org/10.1016/j.scitotenv.2022.158755
8. Ferraro A, Parisi A, Barbone E et al (2023) Characterizing contaminants distribution in marine-coastal sediments through multivariate and nonparametric statistical analyses: a complementary strategy supporting environmental monitoring and control. Environ Monit Assess 195:59. https://doi.org/10.1007/s10661-022-10617-4
9. Birch GF, Taylor SE, Matthai C (2001) Small-scale spatial and temporal variance in the concentration of heavy metals in aquatic sediments: a review and some new concepts. Environ Pollut 113(3):357–372. https://doi.org/10.1016/s0269-7491(00)00182-2
10. El Zokm GM, Ibrahim MIA, Mohamed LA, El-Mamoney M (2020) Critical geochemical insight into Alexandria coast with special reference to diagnostic ratios (TOC/TN & Sr/Ca) and heavy metals ecotoxicological hazards. Egypt J Aquat Res 46:27–33. https://doi.org/10.1016/j.ejar.2019.12.006
11. El Zokm GM, Al-Mur BA, Okbah MA (2020b) Ecological risk indices for heavy metal pollution assessment in marine sediments of Jeddah Coast in the Red Sea. Int J Environ Analyt Chem. https://doi.org/10.1080/03067319.2020.1784888
12. Brito GB, da Silva Júnior JB, Dias LC, de Santana SA, Hadlich GM, Ferreira SLC (2020) Evaluation of the bioavailability of potentially toxic metals in surface sediments collected from a tropical river near an urban area. Mar Pollut Bull 156:111215. https://doi.org/10.1016/j.marpolbul.2020.111215
13. El-Said GF, El-Zokm GM, El Sayed AA, El Ashmawy AA, Shreadah MA (2020) Anomalous fluctuation of halogens in relation to the pollution status along Lake Mariout, Egypt. J Chem:1–20. https://doi.org/10.1155/2020/8102081
14. Wang P, Zhang L, Lin X, Yan J, Zhang P, Zhao B, Zhang C, Yu Y (2020) Spatial distribution, control factors and sources of heavy metal in the surface sediments of Fudu Estuary waters, East Liaodong Bay, China. Mar Pollut Bull 156:111279. ISSN 0025-326X. https://doi.org/10.1016/j.marpolbul.2020.111279
15. Liu D, Wang J, Yu H et al (2021) Evaluating ecological risks and tracking potential factors influencing heavy metals in sediments in an urban river. Environ Sci Eur 33:42. https://doi.org/10.1186/s12302-021-00487-x

16. Perumal K, Antony J, Muthuramalingam S (2021) Heavy metal pollutants and their spatial distribution in surface sediments from Thondi coast, Palk Bay, South India. Environ Sci Euro 33:63. https://doi.org/10.1186/s12302-021-00501-2
17. Wang D-Y, Zhu M-X, Sun C-H, Ma K, Sun W-X, Zhang X-R, Sun Z-L (2021) Geochemistry of iron and sulfur in the Holocene marine sediments under contrasting depositional settings, with caveats for applications of paleoredox proxies. J Mar Syst. https://doi.org/10.1016/j.jmarsys.2021.103572
18. Fakhraee M, Hancisse O, Canfield DE, Crowe SA, Katsev S (2019) Proterozoic seawater sulfate scarcity and the evolution of ocean-atmosphere chemistry. Nat Geosci 12(5):375. https://doi.org/10.1038/s41561-019-0351-5
19. Norman AL, Giesemann A, Krouse HR, Jäger HJ (2002) Sulphur isotope fractionation during sulphur mineralization: results of an incubation-extraction experiment with a Black Forest soil. Soil Biol Biochem 34(10):1425–1438. https://doi.org/10.1016/S0038-0717(02)00086-X
20. Förstner U (1995) Non-linear release of metals from aquatic sediments. In: Salomons W, Stigliani WM (eds) Biogeody-namics of pollutants in soils and sediments. Springer, Berlin, pp 247–307. https://doi.org/10.1007/978-3-642-79418-6_11
21. BoothmanW S, Hansen DJ, Berry WJ, Robson DL, Helmstetter A, Corbin JM et al (2001) Biological response to variation of acid-volatile sulfides and metals in field-exposed spiked sediments. Environ Toxicol Chem 20(2):264–272. https://doi.org/10.1002/etc.5620200206
22. Nizoli EC, Luiz-Silva W (2012) Seasonal AVS–SEM relationship in sediments and potential bioavailability of metals in industrialized estuary, southeastern Brazil. Environ Geochem Health 34:263–272. https://doi.org/10.1007/s10653-011-9430-2
23. Hall LW Jr, Anderson RD (2022) Historical global review of acid-volatile sulfide sediment monitoring data. Soil Syst 6:71. https://doi.org/10.3390/soilsystems6030071
24. Shylesh Chandran MN, Mohan M, Ramasamy EV (2018) Risk assessment of heavy metals in Vemba nad Lake sediments (south-west coast of India), based on acid-volatile sulfide (AVS)-simultaneously extracted metal (SEM) approach. Environ Sci Pollut Res 25:7333–7345.https://doi.org/10.1007/s11356-017-0997-8
25. Rickard D, Morse JW (2005) Acid volatile sulfide (AVS). Mar Chem 97(3–4):141–197. https://doi.org/10.1016/j.marchem.2005.08.004
26. Van den Berg GA, Buykx SEJ, Van den Hoop MAGT, Van der Heijdt LM, Zwolsman JJG (2001) Vertical profiles of trace metals and acid-volatile sulphide in a dynamic sedimentary environment: Lake Ketel, The Netherlands. Appl Geochem 16(7–8):781–791. https://doi.org/10.1016/S0883-2927(00)00076-7
27. Hammerschmidt CR, Burton GA (2010) Measurements of acid volatile sulfide and simultaneously extracted metals are irreproducible among laboratories. Environ Toxicol Chem 29(7):1453–1456. https://doi.org/10.1002/etc.173
28. Queiroz HM, Nóbrega GN, Otero XL, Ferreira TO (2018) Are acid volatile sulfides (AVS) important trace metals sinks in semi-arid mangroves? Mar Pollut Bull 126:318–322. https://doi.org/10.1016/j.marpolbul.2017.11.020
29. Bayen S (2012) Occurrence, bioavailability and toxic effects of trace metals and organic contaminants in mangrove ecosystems: a review. Environ Int 1(48):84–101. https://doi.org/10.1016/j.envint.2012.07.008
30. Cadier C, Bayraktarov E, Piccolo R, Adame MF (2020) Indicators of coastal wetlands restoration success: a systematic review. Front Mar Sci 7:600220. https://doi.org/10.3389/fmars.2020.600220
31. Shen L, Huang T, Chen Y et al (2022) Diverse transformations of sulfur in seabird-affected sediments revealed by microbial and stable isotope analyses. J Ocean Limnol. https://doi.org/10.1007/s00343-021-1173-z
32. De Jonge M, Dreesen F, De Paepe J, Blust R, Bervoets L (2009) Do acid volatile sulfides (AVS) influence the accumulation of sediment-bound metals to benthic invertebrates under natural field conditions? Environ Sci Technol 43(12):4510–4516. https://doi.org/10.1021/es8034945
33. Campana O, Simpson SL, Spadaro DA, Blasco J (2012) Sub-lethal effects of copper to benthic invertebrates explained by sediment properties and dietary exposure. Environ Sci Technol 46:6835–6842. https://doi.org/10.1021/es2045844

34. Zhang C, Yu Z, Zeng G, Jiang M, Yang Z, Cui F, Zhu MY, Shen LQ, Hu L (2014) Effects of sediment geochemical properties on heavy metal bioavailability. Environ Int 73:270–281. https://doi.org/10.1016/j.envint.2014.08.01
35. Chapman PM (1996) Presentation and interpretation of sediment quality triad data. Ecotoxicology 5:327–339. https://doi.org/10.1007/BF00119054
36. Yu KC, Tsai LJ, Chen SH, Ho ST (2001) Chemical binding of heavy metals in anoxic river sediments. Water Res 35:4086–4094. https://doi.org/10.1016/S0043-1354(01)00126-9
37. Fang T, Li X, Zhang G (2005) Acid volatile sulfide and simultaneously extracted metals in the sediment cores of the Pearl River Estuary, South China. Ecotoxicol Environ Safety 61(3):420–431. https://doi.org/10.1016/j.ecoenv.2004.10.004
38. Campana O, Rodriguez A, Blasco J (2009) Identification of a potential toxic hot spot associated with AVS spatial and seasonal variation. Arch Environ Contam Toxicol 56:416–425. https://doi.org/10.1007/s00244-008-9206-6
39. Hernández-Crespo C, Martín M, Ferrís M, Oñate M (2012) Measurement of acid volatile sulphide and simultaneously extracted metals in sediment from Lake Albufera (Valencia, Spain). Soil Sediment Contamin Int J 21(2):176–191. https://doi.org/10.1080/15320383.2012.649374
40. Yin HB, Fan CX, Ding SM, Zhang L, Li B (2008) Acid volatile sulfides and simultaneously extracted metals in a metal-polluted area of Taihu Lake, China. Bull Environ Contam Toxicol 80(4):351–355. https://doi.org/10.1007/s00128-008-9387-8
41. Machado W, Villar LS, Monteiro FF, Viana LCA, Santelli RE (2010) Relation of acid-volatile sulfides (AVS) with metals in sediments from eutrophicated estuaries: is it limited by metal-to-AVS ratios? J Soils Sediments 10:1606–1610. https://doi.org/10.1007/s11368-010-0297-0
42. Liu C, Kong M, Zhang L et al (2020) Metal bioavailability during the periodic drying and rewetting process of littoral anoxic sediment. J Soils Sediments 20:2949–2959. https://doi.org/10.1007/s11368-020-02634-y
43. Aller RC (1977) The influence of macrobenthos on chemical diagenesis of marine sediments. Ph.D. dissertation Thesis, Yale, NewHaven, CT, 600 p
44. Widdel F, Hansen TA (1992) The dissimilatory sulfate- and sulfur reducing bacteria. In: Balows A, Truper HG, Dworkin M, Harder W, Schleifer KH (eds) The prokaryotes. A handbook on the biology of bacteria: ecophysiology, isolation, identification, applications. Springer, New York, pp 583–624. https://doi.org/10.1007/978-1-4757-2191-1?page=3
45. Morse JW (1999) Sulfides in sandy sediments: new insights on the reactions responsible for sedimentary pyrite formation. Aquat Geochem 5:75–85. https://doi.org/10.1023/A:1009620021442
46. Meysman JRP, Middelburg JJ (2005) cid-volatile sulfide (AVS)—a comment. Mar Chem 97:206–212. https://doi.org/10.1016/j.marchem.2005.08.005
47. Tisserand D, Guédron S, Razimbaud S, Findling N, Charlet L (2021) Acid volatile sulfides and simultaneously extracted metals: a new miniaturized 'purge and trap' system for laboratory and field measurements. Talanta 233:22490. https://doi.org/10.1016/j.talanta.2021.122490
48. Chandran MSS, Sudheesh S, Ramasamy EV, Mohan M (2012) Sulphur fractionation in the sediments of cochin estuary. J Environ 1:1–6. https://www.scribd.com/document/200393497/Research-Paper-Sulphur-Fractionation-in-the-Sediments-of-Cochin-Estuary
49. Ju Y-R, Chen C-F, Lim Y, Tsai C-Y, Chen Ch W, Dong Ch D (2022) Developing ecological risk assessment of metals released from sediment based on sediment quality guidelines linking with the properties: a case study for Kaohsiung Harbor. Sci Total Environ 852:158407. https://doi.org/10.1016/j.scitotenv.2022.158407
50. Howarth RW (1984) The ecological significance of sulfur in the energy dynamics of salt marsh and coastal marine sediments. Biogeochemistry 1(1):5–27. https://doi.org/10.1007/BF02181118
51. Thamdrup B, Fossing H, Jørgensen BB (1994) Manganese, iron and sulfur cycling in a coastal marine sediment, Aarhus Bay, Denmark. Geochim Cosmochim Acta 58:5115–5129. https://doi.org/10.1016/0016-7037(94)90298-4

52. Jørgensen BB, Findlay AJ, Pellerin A (2019) The biogeochemical sulfur cycle of marine sediments. Front Microbiol 10:849. https://doi.org/10.3389/fmicb.2019.00849
53. Chen Y, Shen L, Huang T, Chu Z, Xie Z (2020) Transformation of sulfur species in lake sediments at Ardley Island and Fildes Peninsula, King George Island, Antarctic Peninsula. Sci Total Environ 10(703):135591. https://doi.org/10.1016/j.scitotenv.2019.135591
54. Wasmund K, Mußmann M, Loy A (2017) The life sulfuric: microbial ecology of sulfur cycling in marine sediments. Environ Microbiol Rep 9(4):323–344. https://doi.org/10.1111/1758-2229.12538
55. Couture RM, Fischer R, Van Cappellen P et al (2016) Non-steady state diagenesis of organic and inorganic sulfur in lake sediments. United States: N. p. Geochim Cosmochim Acta 194:15–33. https://doi.org/10.1016/j.gca.2016.08.029
56. Luther GW, Findlay AJ, MacDonald DJ, Owings SM, Hanson TE, Beinart RA, Girguis PR (2011) Thermodynamics and kinetics of sulfide oxidation by oxygen: a look at inorganically controlled reactions and biologically mediated processes in the environment. Front Microbiol 2. https://doi.org/10.3389/fmicb.2011.00062
57. Loka Bharathi PA (2008) Sulfur cycle. Encyclopedia Ecol:192–199. https://doi.org/10.1016/b978-0-444-63768-0.00761-7
58. Tang K, Baskaran V, Nemati M (2009) Bacteria of the Sulphur cycle: an overview of microbiology, biokinetics and their role in petroleum and mining industries. Biochem Eng J 44:73–94. https://doi.org/10.1016/j.bej.2008.12.011
59. Glöckner FO, Stal L, Sandaa RA, Gasol JM, O'Gara F, Hernandez F et al (2012) Marine microbial diversity and its role in ecosystem functioning and environmental change. In: McDonough N, Calewaert J-B (eds) Marine board position paper 17 in marine board position paper 17. European Scientific Foundation, Belgium, 84. https://doi.org/10.13140/RG.2.1.5138.6400
60. Postgate JR (1984) The sulphate-reducing bacteria, 2nd ed. Cambridge University Press, Cambridge, UK. ISBN 978-0-521-25791-3
61. Kushkevych I, Hýžová B, Vítˇezová M, Rittmann SK-MR (2021) Microscopic methods for identification of sulfate-reducing bacteria from various habitats. Int J Mol Sci 22:4007. https://doi.org/10.3390/ijms22084007
62. Morse JW, Cornwell JC (1987) Analysis and distribution of iron sulfide minerals in recent anoxic marine sediments. Mar Chem 22:55–69. https://doi.org/10.1016/0304-4203(87)90048-X
63. Di Toro DM, Mahony JD, Hansen DJ, Scott KJ, Hicks MB, Mayr SM, Redmond MS (1990) Toxicity of cadmium in sediments: the role of acid volatile sulfide. Environ Toxicol Chem 9(12):1487–1502. https://doi.org/10.1002/etc.5620091208
64. Jørgensen BB (1990) The sulfur cycle of freshwater sediments: role of thiosulfate. Limnol Oceanogr 35(6):1329–1342. https://doi.org/10.4319/lo.1990.35.6.1329
65. Jørgensen BB, Cohen Y (1977) Solar Lake (Sinai). 5. The sulfur cycle of the benthic cyanobacterial mats. Limnol Oceanogr 22:657–666. https://doi.org/10.4319/lo.1977.22.5.0814
66. Callbeck CM, Canfield DE, Kuypers MMM, Yilmaz P, Lavik G, Thamdrup B, Schubert CJ, Bristow LA (2021) Sulfur cycling in oceanic oxygen minimum zones. Limnol Oceanogr 66(6):2360–2392. https://doi.org/10.1002/lno.11759
67. Middelburg JJ, Levin LA (2009) Coastal hypoxia and sediment biogeochemistry. Biogeosciences 6:1273–1293. https://doi.org/10.5194/bg-6-1273-2009
68. Rabalais NN, Cai W-J, Carstensen DJ, Conley B, Fry X, Hu Z et al (2014) Eutrophication-driven deoxygenation in the coastal ocean. Oceanography 27:172–183. https://doi.org/10.5670/oceanog.2014.21
69. Breitburg D, Levin LA, Oschlies A, Grégoire M, Chavez FP, Conley DJ, Garçon V, Gilbert D, Gutiérrez D, Isensee K, Jacinto GS, Zhang J (2018) Declining oxygen in the global ocean and coastal waters. Science 359(6371):eaam7240. https://doi.org/10.1126/science.aam7240
70. Sampou P, Oviatt CA (1991) Seasonal patterns of sedimentary carbon and anaerobic respiration along a simulated eutrophication gradient. Mar Ecol Prog Ser 72:271–282. https://doi.org/10.3354/meps072271

71. Canfield DE, Stewart FJ, Thamdrup B, De Brabandere L, Dalsgaard T, Delong EF, Revsbech NP, Ulloa O (2010) A cryptic sulfur cycle in oxygen-minimum-zone waters off the Chilean coast. Science 330:1375–1378. https://doi.org/10.1126/science.119688
72. Holmkvist L, Ferdelman TG, Jørgensen BB (2011) A cryptic sulfur cycle driven by iron in the methane zone of marine sediment (Aarhus Bay, Denmark). Geochim Cosmochim Acta 75:3581–3599. https://doi.org/10.1016/j.gca.2011.03.033
73. Böttcher ME (2011) Sulfur cycle. In: Reitner J, Thiel V (eds) Encyclopedia of geobiology. Encyclopedia of earth sciences series. Springer, Dordrecht. https://doi.org/10.1007/978-1-4020-9212-1_200
74. Jasińska A, Burska D, Bolałek J (2012) Sulfur in the marine environment. Oceanol Hydrobiol Stud 41(2):72–82. https://doi.org/10.2478/s13545-012-0019-x
75. Jackson R, Gabric A (2022) Climate change impacts on the marine cycling of biogenic sulfur: a review. Microorganisms 10:1581. https://doi.org/10.3390/microorganisms10081581.a
76. Soliman NF, El Zokm GM, Okbah MA (2018) Risk assessment and chemical fractionation of selected elements in surface sediments from Lake Qarun, Egypt using modified BCR technique. Chemosphere 191:262–271. https://doi.org/10.1016/j.chemosphere.2017.10.049
77. El Zokm GM, El-Said GF, El Ashmawy AA (2022) A comparative study, distribution, predicted no-effect concentration (PNEC) and contamination assessment of phenol with heavy metal contents in two coastal areas on the Egyptian Mediterranean sea coast. Mar Environ Res 179:105687
78. El Zokm GH, Masoud MS, El-Shorbagi EK, Elsamra RMI, Okbah MA (2023) Reactive sulfide dynamic models for predicting metal hazardous in sediments of two northern Egyptian lakes. Mar Pollut Bull 188:114694. https://doi.org/10.1016/j.marpolbul.2023.114694

Chapter 2
Chemistry of Sulfur Components and Factors Controlling AVS Concentrations in Marine Environment

The chemical concept for AVS-SEM models is the reaction of divalent metals displaced iron in FeS rapidly to build more MeS as described in Eq. 2.1.

Outside of the synthesis of H_2S and a few common sulfur forms, there is a knowledge gap in sulfur chemistry [1].

2.1 The Chemistry of Sulfides in Aqueous Solutions

H_2S and HS^- with minor S^{2-} are the main forms of free sulfide S(–II). H_2S has a dual nature (Lewis base or acid), however HS^- acts as a Lewis base. The chemistry of these forms in aquatic environment has been subject to some reasonable restrictions. At acidic pH, H_2S dominates the system, whereas HS^- dominates at alkaline pH. In seawater, [5–25 °C and salinities (5–40)], the value of pK1(H_2S) in seawater, pK1 * (H_2S), can be described as a function of T (in kelvin), and salinity [2].

$$\mathrm{pK1} * (\mathrm{H_2S\ in\ seawater}) = \mathrm{pK1} + \mathrm{AS}^{1/2} + \mathrm{BS} \tag{2.1}$$

$$\mathrm{pK1} = -98.080 + 5765.4/\mathrm{T} + 15.0455 \ln \mathrm{T} \tag{2.2}$$

$$\mathrm{A} = -0.1498 \text{ and } \mathrm{B} = 0.0119.$$

Although less strictly limited, the pK2 (H_2S) is believed to be greater than 18.

G. M. El Zokm, *Ecological Quality Status of Marine Environment*, Earth and Environmental Sciences Library, https://doi.org/10.1007/978-3-031-29203-3_2

2.2 An Overview of Potency AVS Components of Sediments

Deep insight into $FeS + MnS + H_2S + HS^-$ as dissolved components of AVS is necessary for better recognition of the biogeochemistry of sulfide in the aquatic environment. The use of HCl will lead to the evolving of H_2S from dissolved components and iron sulfide minerals, which could be classified as follows:

(1) dissolved iron and sulfur components, as well as their complexes
(2) sulfide clusters FeS_{aq}.

Molecular complexes or nanoparticles with associations of molecules small enough to act as dissolved species are known as aqueous metal sulfide clusters. Clusters of aqueous iron (II) sulfide contribute significantly to the AVS component of anoxic sediments. From Fe_2S_2 up to $Fe_{150}S_{150}$ molecules, the FeS clusters create a series of species. In sulfidic sediments in the marine environment, FeS clusters carry large amounts of Fe. Because H_2S is harmful to aquatic life, the presence of FeS clusters has biogeochemical consequences. By eliminating a part of free aqueous S, the creation of FeS clusters lowers this toxicity (–II) [3]. Since, the p$K1$ for H_2S is around 7, sulfur components of AVS are assumed to be H_2S (aq) and HS^- as major dissolved depending on pH. As pK_2 for H_2S is high, the aqueous S^{2-} ion is a minor contributor to AVS in the aquatic environment. Sulfuroxy anions and polysulfides are non-significant species in the total sulfide budget. Moreover, pyrite formation mainly includes FeS clusters Fe/MnOx [4]. The stoichiometry of these clusters is one-to-one [5] and ranged from simply dissolved clusters to nanoparticles (Figs. 2.1 and 2.2). The equations represented AVS in sediments are being considered on the basis of previous assumption [6, 7]:

$$C_{AVS} = \sum C_{S(-II)\,\text{dissolved}} + \sum C_{S(-II)\,\text{minerals}} + \sum C_{S(-II)\,\text{clusters and nanoparticles}} \tag{2.3}$$

$$\sum C_{S(-II)\,\text{dissolved}} = C_{H_2S(aq)} + C_{HS} - + C_{Sx} + C_{FeHS} \tag{2.4}$$

$$\sum C_{H_2S,\text{mineral}} = C_{\sum FeS} + C_{f FeS_2} \tag{2.5}$$

where S_x; oxidized dissolved sulfur species that produce H_2S when exposed to HCl, and $FeHS_x$; ferrous sulfide complexes. FeS is the phase that reacts with HCl to create H_2S, such as mackinawite and greigite, while FeS_2 is the pyrite proportion that dissolves by HCl.

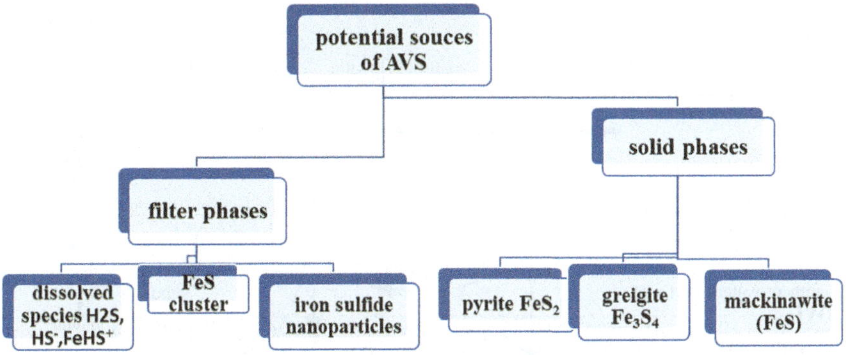

Fig. 2.1 Proposed AVS potential sources (modified from [7])

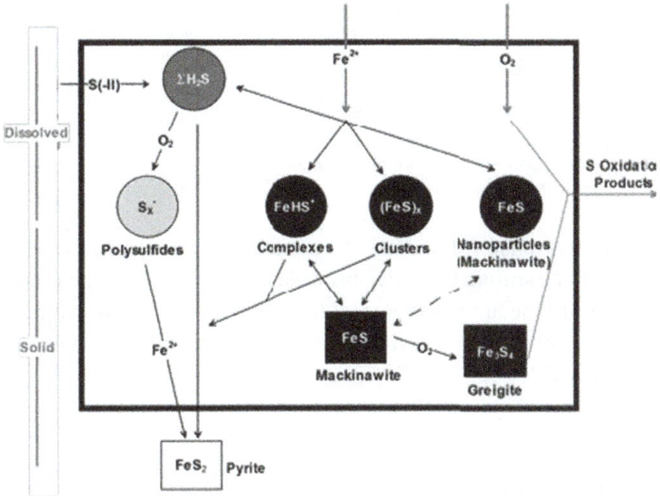

Fig. 2.2 A complex box model for interactions among AVS components [7, 8]

2.3 Factors that Affect AVS Concentrations and Their Mode of Action

The rate of generation of S(–II) and consumption by oxidation or diffusion are the primary drivers of acid volatile sulfide in aquatic sediments. AVS concentrations fluctuated depending on parameters such as organic matter availability, sulfate reduction rate, and sediment redox potential [9]. Age, pH, eating, and the activity of benthic organisms in sediment all impact metal mobility and distribution. These factors regularly interact with one another [10]. Figure 2.3 depicts the formation of AVS and its interaction with heavy metals.

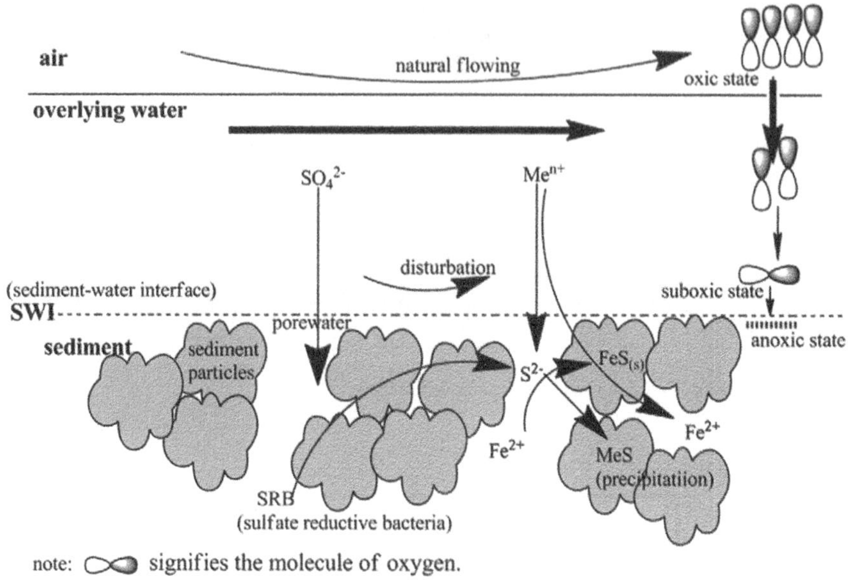

Fig. 2.3 The mode of action of AVS with heavy metals [11]

It is based on the transfer of sulfate- from bottom water to pore water, as well as its destiny through the sulfate reductive bacteria metabolism through pore water with ions of heavy metal. The arrows represent the materials' movement direction, while the thickness of the arrows reflects the number of materials. The n-valence metal ion is represented by Me^{n+}.

2.3.1 Sulfate Reduction Rate

The elimenation of inorganic sulfur-containing compounds from the environment is mostly performed by two types of photoautotrophic (green and purple sulfur-oxidizing bacteria) and by colorless sulfur-oxidizing bacteria trough chemolithotrophic pathway.

Sulfate-reducing bacteria can use molecular hydrogen (H_2) and organic compounds to produce electrons (Fig. 2.4).

$$SO_4{}^{2-} + 8H^+ + 8e^- + Fe^{2+} \xrightarrow{SRB} FeS + 4H_2O \tag{2.6}$$

In the presence of Fe^{2+}, sulfate reduction may result in the production of weakly crystalline FeS [12]. According to the previous equation, AVS mainly as FeS that is poorly crystalline and thermodynamically labile originated due to sufficient sulfate

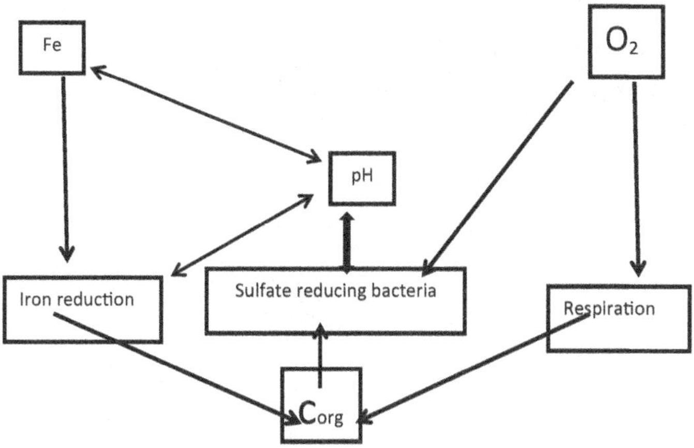

Fig. 2.4 Regulation of sulfate reduction (modified from [15])

supply and under moderately to strongly reducing environmental conditions in sediment with SRB [13, 14].

2.3.1.1 Sulfate Activation

Temporal variation in the sulfate-reducing pathways has been proven in previous research to be rather sensitive [16, 17]. Before sulfate is reduced, it is transferred to bacterial cells and converted to active by the action of the enzyme sulfurylase; ATP, forming adenosine 5′-phosphosulfate and PPi, pyrophosphate; APS as shown in (Figs. 2.5 and 2.6). The reaction is also reversible, and therefore, ATP can be formed from PPi and APS [18, 19].

$$SO_4^{2-} + ATP + 2H^+ = APS + PPi \quad \Delta G^0 = -46\,kJ/mol$$

Further, adenylyl sulfate reductase reduces sulfite, which is then reduced to hydrogen sulfide by dissimilatory sulfite reductases [20]. The conversion of APS to MP necessitates the use of two electrons. In both prokaryotic and eukaryotic cells, this is an important step in the sulfate absorption and dissimilation process. Sulfite reductase converts sulfite to sulfide without producing any intermediates. In the dissimilatory sulfate reduction, this reaction is also a final step. Three important enzymes catalyze the sulfate-reduction mechanism in SRB: ATP sulfurylase, APS reductase, and sulfite reductase. The activation of sulfate by the ATP sulfurylase enzyme is followed by reducting of APS to AMP and sulfite by adenylyl sulfate reductase, followed by reducting to H_2S by dissimilatory sulfite reductases. The conversion of APS to AMP necessitates the use of two electrons and this is a crucial stage [14].

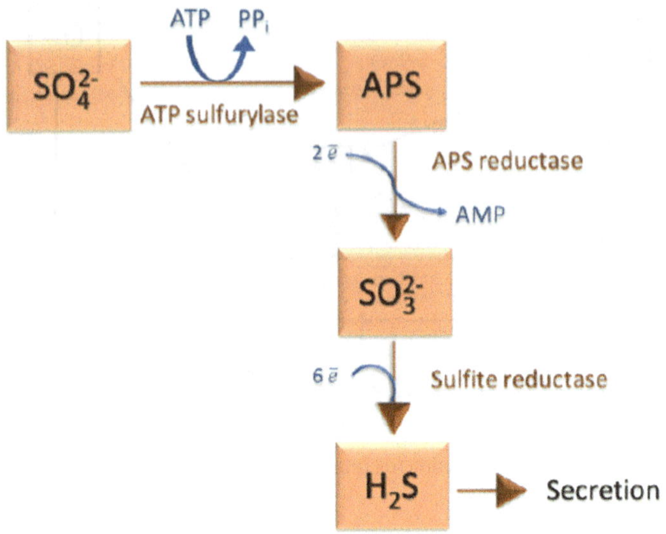

Fig. 2.5 Scheme of the dissimilatory sulfate reduction [14]

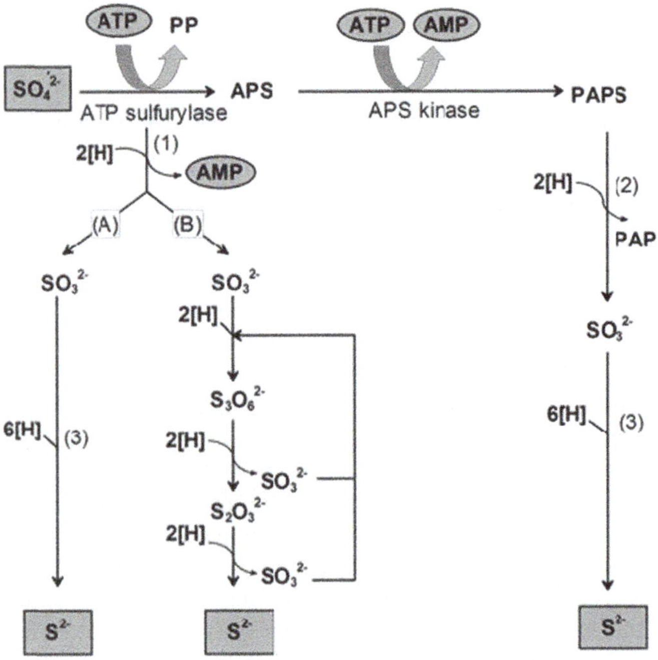

Fig. 2.6 Dissimilative and assimilative sulfate reduction Pathways [8]

2.3.2 *Oxidation–Reduction Chemistry of AVS During Analysis*

The various AVS-S extraction methods have changed over time, making it challenging to compare the outcomes. Oxidation-reduction potential (Eh; mV) is a critical element that determines electron availability and allows for predicting heavy metal stability and bioavailability in sediment. Sediment redox zones can be classified into three levels vertically: oxygen reduction (oxic), nitrate, iron and manganese reduction (suboxic), and sulfate reduction and methanogenesis (anoxic) [21–26]. To evaluate metal bioavailability, discrimination between these layers is a must. When the anaerobic redox state is overcome by aerobic in sediment, such as during floods, metal bioavailability ranges from high to low. AVS in sediments is strongly related to Eh as well as water content [27].

Higher AVS levels are in sediments with low Eh and high-water content. Small changes in redox conditions result in large changes in pyrite super saturation in suboxic systems near the SO_4^{2-}/S(–II) boundary (Figs. 2.7 and 2.8). This suggests that the rate of origination of pyrite is highly dependent on local physicochemical conditions. Pyrite formation will be kinetically inhibited in some places, allowing other AVS components to take precedence; but, in surrounding regions with slightly differing redox conditions, pyrite formation is rapid, allowing pyrite to dominate the system [6, 21].

The susceptibility of certain sediment acid-volatile sulfide (AVS) components to chemical oxidation is a restricting factor in measuring AVS accurately in sediment samples. Because of this well-documented sensitive oxidation, the analytical technique developed for AVS required oxygen-free conditions. The AVS stability in sediment samples limits the accuracy of the SEM and AVS results. Changes in

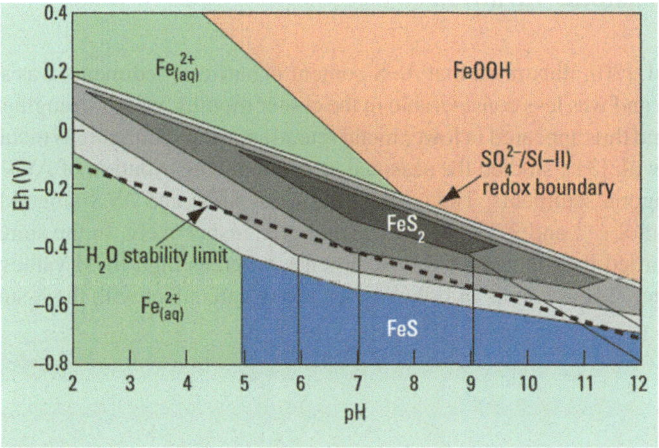

Fig. 2.7 pH–Eh diagram at 25 °C for the iron–sulfide–water system [6]

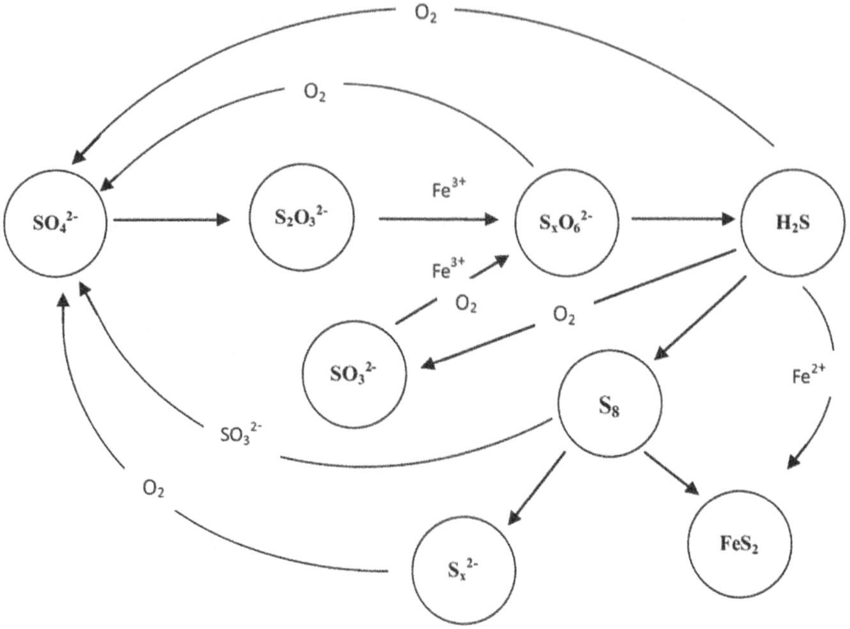

Fig. 2.8 Oxidation reduction reactions of sulfur species [1]

AVS content during monitoring, especially due to oxidation, might result in incorrect SEM: AVS ratios and toxicological categorization of sediments [28, 29].

2.3.3 *Seasonal Variation*

Harris et al. [30], illustrated that AVS content in surficial sediments was seasonally dependent and was less considerable in the cooler months, which strengthened metal mobility and thus appeared in lower metal retention in the sediments as metal sulfides. El Zokm et al. [31], studied the seasonal effect on the distribution of AVS and SEM in Edku lagoon sediments. The concentrations of SEM and AVS ranged from 0.86 to 3.82 μ mol g^{-1} and 2.53 to 138.20 μmol g^{-1}, respectively. In the summer, AVS levels recorded the highest level in sediments, whereas high SEM values achieved in the winter. These results in Edku lagoon are in agreement with the assumption of [30].

2.4 Mechanisms Controlling Sulfide Production in Marine Environment Highlighted Mn and Fe Roles

The only significant biogeochemical activity that produces sulfate on Earth is metal sulfide oxidation. Processing ores for metal recovery sometimes involves using metal sulfides that develop and oxidize in sediments. Mechanisms describe MeS oxidation may be influenced by environment status as pH, oxidants and surface-controlled processes [32], structure of MeS, impurities, dislocation lines; linear defects in crystals and stacking faults; planar defects [33], photochemical reactions [34]. Microorganisms are needed in oxidation of transitional sulfur compounds produced by the chemical dissociation of MeS since, transitional sulfur species are oxidized either chemically by MnO_2 or by microorganisms using an electron acceptor (nitrate) in anoxic and pH-neutral environments. According to [35], the principal quantitative oxidants for sulfide in subsurface sediments are Fe/MnOx minerals. FeOx are weaker oxidants than MnOx and react more slowly as a result. Sulfide oxidation by Mn^{4+} involves a two-electron transfer with S^0 as an immediate. Böttcher and Thamdrup [37]. Genus *Sulfurimonas* of Bacteria has been shown autotrophically growing during oxidizing S^{2-} to $SO_4{}^{2-}$ with MnO_2 [38, 39]. In biodegradable composite near-surface sediments with feedback of O_2 and NO_3, as well as the presence of metal oxides, sulfide oxidation goes in all paths to sulfate, thus closing the sulfur cycle. The majority of the oxidant, Fe (III), is buried in subterranean sediments, and only partial oxidation may occur, resulting in the formation of intermediates such iron mono sulfide (FeS), pyrite (FeS_2), and elemental sulfur (S^0). Sulfide in anoxic sea sediments reacts at significantly different speeds depending on the crystalline structure and reactive surface area of the metal oxides.

2.4.1 *Sulfide Oxidation Mechanisms*

Two mechanisms; polysulfide and the thiosulfate are used to explain the oxidation pathway of metal sulfide [35, 40, 41]. The polysulfide mechanism (Fig. 2.9) states that metal sulfides that dissolve in acid oxidized to elemental sulfur by a series of electron transfers, $Fe^{3+}(H_2O)n$ oxidize the S of FeS_2 to a RSO_2 (OH) (sulfonic acid group), according to the thiosulfate process. As a result of this transition, the attack of $Fe^{3+}(H_2O)_6$ breaks the bonds between the Fe atom and the two sulfur atoms in pyrite, resulting in the creation of hydrated Fe^{2+} and thiosulfate, which are then physiologically oxidized by bacteria that are moderately acidophilic and oxidize sulfur compounds. In thiosulfate mechanism (Fig. 2.10), Fe^{3+}, NO^{3-}, MnO_2, or O_2 are used as oxidants in the oxidation of sulfur intermediate species. The cycle's initial intermediary sulfur molecule, are broken down into sulfate via tetrathionate, disulfane, monosufonic acid, and trithionate. Sulfate, pentathionate, and elemental sulfur are produced in side reactions.

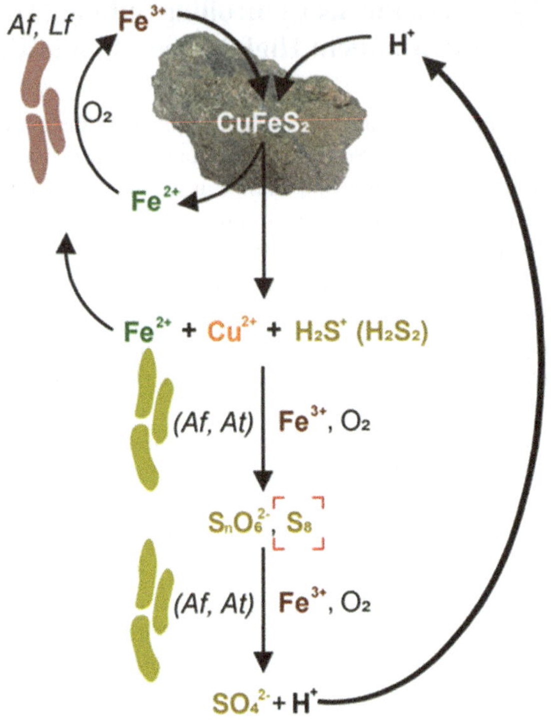

Fig. 2.9 A scheme of the polysulfide mechanism for acid soluble metal sulfides [41] ***Af***; *At. ferrooxidans* from Iron (II)-oxidizing bacteria, ***Lf***; *L. ferrooxidans* from Iron(II)-oxidizing bacteria

In conclusion, the polysulfide mechanism is primarily mediated by thiosulfate, whereas the thiosulfate mechanism is primarily mediated by polythionates. Depending on the intermediate creation of elemental sulfur, polysulfides, and ferrous iron minerals, several pyrite formation paths may be followed [36]. In marine sediments, two reactions appear to predominate Eqs. 2.7 [42] and 2.8 [2]:

$$H_2S\,\text{pathway: } FeS + H_2S \rightarrow FeS_2 + H_2(5.5) \tag{2.7}$$

$$\text{Polysulfide pathway: } FeS + S_x^{2-} \rightarrow FeS_2 + S_{x-1}^{2-} \tag{2.8}$$

The following stoichiometries are obtained in the presence of Fe/MnOx [43]:

$$\begin{aligned} &3S^{\circ} + FeOOH \Rightarrow SO_4{}^{2-} + 2FeS + 2H \\ &\quad + 3S + FeOOH \rightarrow SO_4{}^{2-} + 2FeS + 2H^+ \end{aligned} \tag{2.9}$$

$$S^{\circ} + 3MnO_2 + 4H^+ \Rightarrow SO_4{}^{2-} + 3Mn^{2+} + 2H_2O \tag{2.10}$$

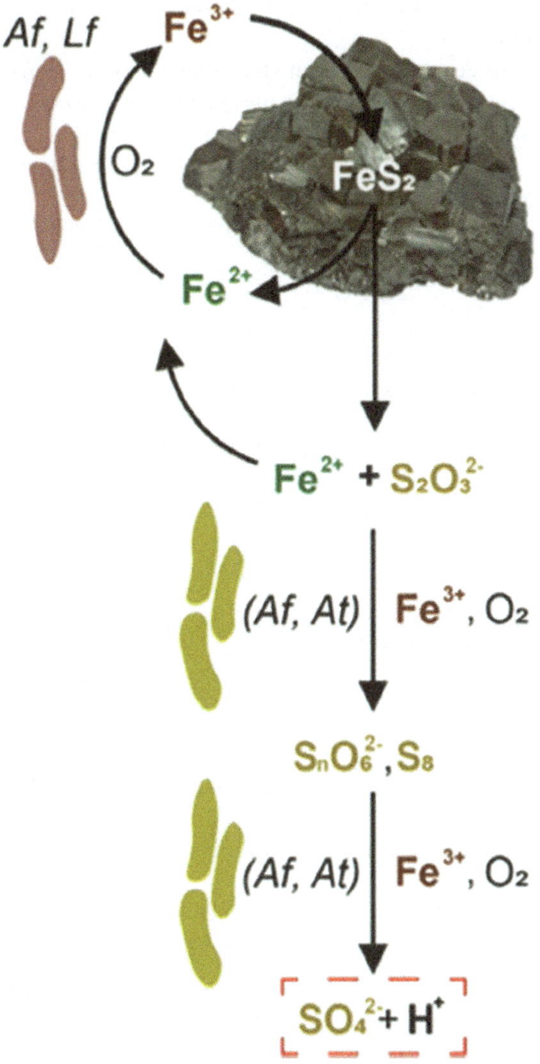

Fig. 2.10 Thiosulfate mechanism of pyrite oxidation [41]

Meysman and Middelburg [44], highlighted the chemistry of AVS focusing on the following points:

1. AVS is an important and active chemical constituent.
2. How can microorganisms affect the chemistry of AVS?
3. The pathways of the sulfur cycle
4. The reaction mechanisms
5. Dissolved fractions in AVS
6. Digenetic modeling governing AVS chemistry.

In diagenetic models, the sulfur cycle contains a subset of the processes listed below.

To get a consistent model formulation, the diagenetic models governing sulfur chemistry must be built on the assumption of accommodation between model species suited for the reactions and reaction equations.

The sedimentary sulfur cycle begins with sulfate reduction and the production of iron sulfide.

Hydrogen sulfide species are generated first by sulfate-reducing bacteria.

$$2CH_2O + 1/2SO_4{}^{2-} + H^+ \rightarrow 2CO_2 + 1/2H_2S + 2H_2O \qquad (2.11)$$

H_2S is the most reduced, is a weak acid.

Fraction of sulfide is re-oxidized by iron oxides and give sulfur

$$FeOOH + 1/2H_2S + 2H^+ \longrightarrow 1/2S^0 + Fe^{2+} + 2H_2O \qquad (2.12)$$

Formation of FeS

$$Fe^{2+} + H_2S \longrightarrow FeS + 2H^+ \qquad (2.13)$$

Formation of pyrite

$$FeS + S^0 \longrightarrow FeS_2 \qquad (2.14)$$

$$FeS + H_2S \longrightarrow FeS_2 + H_2 \qquad (2.15)$$

Furthermore, reduced sulfur compounds can be re-oxidized with oxygen supply.

$$H_2S + 2O_{2(aq)} \longrightarrow SO_4{}^{2-} + 2H^+ \qquad (2.16)$$

$$S^0 + 3/2O_2(aq) + H_2O \longrightarrow SO_4{}^{2-} + 2H^+ \qquad (2.17)$$

$$FeS + 2O_2(aq) \longrightarrow Fe^{2+} + SO_4{}^{2-} \qquad (2.18)$$

$$FeS_2 + 7/2O_2(aq) + H_2O \longrightarrow Fe^{2+} + 2SO_4{}^{2-} + 2H^+ \qquad (2.19)$$

Equations (2.11–2.19) describe the sulfur cycle in diagenetic models. Some notes must be taken into consideration; the fate of elemental sulfur evaluated from Eq. 2.12 is not followed [45, 46]. The fate of S^0 is unrealistic for actual sediments. But, Eq. 2.14 describes the removal of elemental sulfur through the formation of pyrite. Also, Eq. 2.17 describes the re-oxidation of S^0 to produce sulfate. Equation 2.14 illustrates the formulation of pyrite but does not include a diagenetic equation for hydrogen [47]. Hydrogen accumulation is improbable since hydrogen is highly reactive and quantities in natural sediments are generally in the nanomolar range. A hydrogen removal mechanism must be incorporated to close the mass balance [48].

References

1. Basafa M, Hawboldt K (2019) Reservoir souring: sulfur chemistry in offshore oil and gas reservoir fluids. J Petrol Explor Product Technol 9:1105–1118. https://doi.org/10.1007/s13202-018-0528-2
2. Rickard D, Luther GW (2007) Chemistry of iron sulfides. Chem Rev 107:514–562. https://doi.org/10.1021/cr0503658
3. Luther G, Rozan T, Taillefert M et al (2001) Chemical speciation drives hydrothermal vent ecology. Nature 410:813–816. https://doi.org/10.1038/35071069
4. Rickard D, Oldroyd A, Cramp A (1999) Voltametric evidence for soluble FeS complexes in anoxic estuarine muds. Estuaries 22:693–701. https://doi.org/10.2307/1353056
5. Theberge SM, Luther GW III (1997) Determination of the electrochemical properties of asoluble aqueous FeS cluster present in sulfidic systems. Aquat Geochem 3:191–211. https://doi.org/10.1023/a:1009648026806
6. Morse JW, Rickard D (2004) Chemical dynamics of sedimentary acid volatile sulfide. Environ Sci Technol 38:131A-136A. https://doi.org/10.1021/es040447y
7. Rickard D, Morse JW (2005) Acid volatile sulfide (AVS). Mar Chem 97(3–4):141–197. https://doi.org/10.1016/j.marchem.2005.08.004
8. Jasińska A, Burska D, Bolałek J (2012) Sulfur in the marine environment. Oceanol Hydrobiol Stud 41(2):72–82. https://doi.org/10.2478/s13545-012-0019-x
9. Van Griethuysen C, Luitwielera M, Joziasseb J, Koelmansa AA (2005) Temporal variation of trace metal geochemistry in floodplain lake sediment subject to dynamic hydrological conditions. Environ Pollut 137(2):281–294. https://doi.org/10.1016/j.envpol.2005.01.023
10. Burton GA, Green A, Baudo R, Forbes V, Nguyen LT, Janssen CR, Kukkonen J, Leppanen M, Maltby L, Soares A, Kapo K, Smith P, Dunning J (2007) Characterizing sediment acid volatile sulfide concentrations in European streams. Environ Toxicol Chem 26(1):1–12. https://doi.org/10.1897/05-708r.1
11. Zhang C, Yu Z, Zeng G, Jiang M, Yang Z, Cui F, Zhu MY, Shen LQ, Hu L (2014) Effects of sediment geochemical properties on heavy metal bioavailability. Environ Int 73:270–281. https://doi.org/10.1016/j.envint.2014.08.01
12. Burton ED, Bush RT, Sullivan LA (2006) Reduced inorganic sulfur speciation in drain sediments from acid sulfate soil landscapes. Environ Sci Technol 40(3):888–893. https://doi.org/10.1021/es0516763
13. Van Griethuysen C, De Lange H, Van den Heuij M, De Bies S, Gillissen F, Koelmans A (2006) Temporal dynamics of AVS and SEM in sediment of shallow freshwater floodplain lakes. Appl Geochem 21:632–642. https://doi.org/10.1016/j.apgeochem.2005.12.010
14. Kushkevych I, Hýžová B, Vítˇezová M, Rittmann SK-MR (2021) Microscopic methods for identification of sulfate reducing bacteria from various habitats. Int J Mol Sci 22:4007. https://doi.org/10.3390/ijms22084007
15. Koschorreck M, Wendt-Potthoff K (2012) A sediment exchange experiment to assess the limiting factors of microbial sulfate reduction in acidic mine pit lakes. J Soils Sediments 12:1615–1622. https://doi.org/10.1007/s11368-012-0547-4
16. Leonard EN, Cotter AM, Ankley GT (1996) Modified diffusion method for analysis of acid volatile sulfides and simultaneously extracted metals in freshwater sediment. Environ Toxicol Chem 15(9):1479–1481. https://doi.org/10.1002/etc.5620150908
17. Kristensen E, Bouillon S, Dittmar T, Marchand C (2008) Organic carbon dynamics in mangrove ecosystems: a review. Aquat Bot 89:201–219. https://doi.org/10.1016/j.aquabot.2007.12.00

18. Kushkevych IV, Antonyak HL, Bartoš M (2014) Kinetic properties of dissimilatory adenosine tri-phosphate sulfurylase of intestinal sulfate-reducing bacteria. Ukrainian Biochem J 86(6):129–138. https://doi.org/10.15407/ubj86.06.129
19. Kushkevych I, Abdulina D, Kováč J, Dordević D, Vítězová M, Iutynska G, Rittmann SK-MR (2020) Adenosine-5′-phosphosulfate- and sulfite reductases activities of sulfate-reducing bacteria from various environments. Biomolecules 10:921. https://doi.org/10.3390/biom10060921
20. Postgate J (1959) Sulphate reduction by bacteria. Annu Rev Microbiol 13:505–520. https://doi.org/10.1146/annurev.mi.13.100159.002445
21. Butler I, Rickard D (2000) Framboidal pyrite formation via the oxidation of iron (II) monosulfide by hydrogen sulphide. Geochim Cosmochim Acta 64:2665–2672. https://doi.org/10.1016/S0016-7037(00)00387-2
22. Gonzalez AM (2002) Oxidation chemistry of acid-volatile sulfide during analysis. Environ Toxicol Chem 21(5):980–983. https://doi.org/10.1002/etc.5620210512
23. Di Toro DM, Mcgrath JA, Hansen DJ, Berry WJ, Paquin PR, Mathew R, Wu KB, Santore RC (2005) Predicting sediment metal toxicity using a sediment biotic ligand model: methodology and initial application. Environ Toxicol Chem 24(10):2410. https://doi.org/10.1897/04-413r.1
24. Kelderman P, Osman AA (2007) Effect of redox potential on heavy metal binding forms in polluted canal sediments in Delft (The Netherlands). Water Res 41(18):4251–4261. https://doi.org/10.1016/j.watres.2007.05.05
25. Luther GW, Findlay AJ, MacDonald DJ, Owings SM, Hanson TE, Beinart RA, Girguis PR (2011) Thermodynamics and kinetics of sulfide oxidation by oxygen: a look at inorganically controlled reactions and biologically mediated processes in the environment. Front Microbiol 2. https://doi.org/10.3389/fmicb.2011.00062
26. Strom D, Simpson SL, Batley GE, Jolley DF (2011) The influence of sediment particle size and organic carbon on toxicity of copper to benthic invertebrates in oxic/suboxic surface sediments. Environ Toxicol Chem 30:1599–1610. https://doi.org/10.1002/etc.531
27. Mackey AP, Mackay S (1996) Spatial distribution of acid-volatile sulphide concentration and metal bioavailability in man-grove sediments from the Brisbane River, Australia. Environ Pollut 93(2):205–209. https://doi.org/10.1016/0269-7491(96)00031-0
28. USEPA (1994) Methods for measuring the toxicity and bioaccumulation of sediment-associated contaminants with freshwater invertebrates. 600/R-94/024. Washington, DC
29. Becker DC, Ginn TC (1995) Effects of storage time on toxicity of sediments from Puget Sound, Washington. Environ Toxicol Chem 14:829–835. https://doi.org/10.1002/etc.5620140513
30. Harris S, Xu X, Mills G (2020) Metal-sulfide dynamics in a constructed wetland in the Southeastern United States. Wetlands Ecol Manage 28:847–861. https://doi.org/10.1007/s11273-020-09749-6
31. El Zokm GM, Okbah MA, Younis AM (2015) Assessment of heavy metals pollution using AVS-SEM and fractionation techniques in Edku Lagoon sediments, Mediterranean Sea. J Environ Sci Health Part A Toxic/Hazard Subst Environ Eng 50:571–584. https://doi.org/10.1080/10934529.2015.994945
32. Todd EC, Sherman DM, Purton JA (2003) Surface oxidation of pyrite under ambient atmospheric and aqueous (pH = 2 to 10) conditions: electronic structure and mineralogy from X-ray absorption spectroscopy. Geochimica et Cosmochimica Acta 67:881–893. https://doi.org/10.1016/S0016-7037(02)00957-2
33. Thomas JE, Skinner WM, Smart RSC (2003) A comparison of the dissolution behavior of troilite with other iron (II) sulfides; implications of structure. Geochimica et Cosmochimica Acta 67:831–843. https://doi.org/10.1016/S0016-7037(02)01146-8
34. Giannetti BF, Bonilla SH, Zinola CF, Rabóczkay T (2001) A study of the main oxidation products of natural pyrite by voltammetric and photoelectrochemical responses. Hydrometallurgy 60:41–53.https://doi.org/10.1016/S0304-386X(00)00158-4
35. Findlay AJ, Pellerin A, Laufer K, Jørgensen BB (2020) Quantification of sulphide oxidation rates in marine sediment. Geochim Cosmochim Acta 280:441–452. https://doi.org/10.1016/j.gca.2020.04.007

36. Findlay A.J, (2016) Microbial impact on polysulfide dynamics in the environment, FEMS Microbiology Letters, 363(1), fnw103, https://doi.org/10.1093/femsle/fnw103
37. Böttcher ME, Thamdrup B (2001) Anaerobic sulfide oxidation and stable isotope fractionation associated with bacterial sulfur disproportionation in the presence of MnO_2. Geochim Cosmochim Acta 65:1573–1581. https://doi.org/10.1016/S0016-7037(00)00622-
38. Jiang M, Sheng Y, Liu Q, Wang W, Liu X (2021) Conversion mechanisms between organic sulfur and inorganic sulfur in surface sediments in coastal rivers. Sci Total Environ 752. https://doi.org/10.1016/j.scitotenv.2020.141829
39. Henkel JV, Dellwig O, Pollehne F, Herlemann DPR, Leipe T, Schulz-Vogt HN (2019) A bacterial isolate from the Black Sea oxidizes sulfide with manganese (IV) oxide. Proc Natl Acad Sci USA 116:12153–12155. https://doi.org/10.1073/pnas.1906000116
40. Schippers A, Sand W (1999) Bacterial leaching of metal sulfides proceeds by two indirect mechanisms via thiosulfate or via polysulfides and sulfur. Appl Environ Microbiol 1999(65):319–321. https://doi.org/10.1128/AEM.65.1.319-321
41. Vera M, Schippers A, Hedrich S et al (2022) Progress in bioleaching: fundamentals and mechanisms of microbial metal sulfide oxidation—part A. Appl Microbiol Biotechnol 106:6933–6952. https://doi.org/10.1007/s00253-022-12168-7
42. Thiel J, Byrne JM, Kappler A, Schink B, Pester M (2019) Pyrite formation from FeS and H_2S is mediated through microbial redox activity. Proc Natl Acad Sci USA 116:6897–6902. https://doi.org/10.1073/pnas.1814412116
43. Böttcher ME (2011) Sulfur cycle. In: Reitner J, Thiel V (eds) Encyclopedia of geobiology. Encyclopedia of earth sciences series. Springer, Dordrecht. https://doi.org/10.1007/978-1-4020-9212-1_200
44. Meysman JRP, Middelburg JJ (2005) Acid-volatile sulfide (AVS)—a comment. Mar Chem 97:206–212. https://doi.org/10.1016/j.marchem.2005.08.005
45. Wang YF, Van Cappellen P (1996) A multicomponent reactive transport model of early diagenesis: application to redox cycling in coastal marine sediments. Geochim Cosmochim Acta 60:2993–3014. https://doi.org/10.1016/0016-7037(96)00140-8
46. Furukawa Y, Smith AC, Kostka JE, Watkins J, Alexander CR (2004) Quantification of macrobenthic effects on diagenesis using a multicomponent inverse model in salt marsh sediments. Limnol Oceanogr 49:2058–2072. https://doi.org/10.4319/lo.2004.49.6.2058
47. Wijsman JWM, Herman PMJ, Middelburg JJ, Soetaert K (2002) A model for early diagenetic processes in sediments of the continental shelf of the Black Sea. Estuar Coast Shelf Sci 54:403–421. https://doi.org/10.1006/ecss.2000.0655
48. Meysman FJR, Middelburg JJ, Herman PMJ, Heip CHR (2003) Reactive transport in surface sediments: II. Media: an object-oriented problem-solving environment for early diagenesis. Comput Geosci 29:301–318. https://doi.org/10.1016/S0098-3004(03)00007-4

Chapter 3
Experimental Approach to Sampling, Storage, Extraction, Determination of AVS-SEM

3.1 Sampling Technique, Sediment Storage and Pretreatment

The AVS technique measures metals that are bound to sulfide fraction as well as a set of dissolved sulfur species or nano particulates present in sediment pore water and amorphous crystallized sulfides such as mackinawite (FeS), greigite (Fe_3S_4), pyrite (FeS_2), and manganese sulfides. However, organic sulfides' role in AVS is still unclear [1–5].

Before AVS extraction, the major objective of sample preservation and preparation is keeping away from redox state change. Some researchers can examine fresh sediments, whereas others had to freeze or store them at 4 °C before doing so [1, 6–8]. The optimum conditions for processing sediment, according to Lasorsa and Casas [9], were to manage it in the presence of nitrogen, keep it at 4 °C or freeze it, and analyze it within 15 days. Billon [10], found that fresh or dried sediment under N_2 yielded the highest AVS recovery, but that drying sediment under O_2 resulted in a 94% loss.

The sediment's physico-chemical characteristics alter because of its storage. The toxicity of sediment can be influenced by temperature and storage time. De Lange [4], found a substantial rise in AVS when sediment was held at 4 °C due to the activity of sulfate-reducing bacteria during storage. However, after storing hazardous field sediments at 4 °C for a lengthy time, other research found no change in biological reactions [11].

Several investigations indicated that storage affected sediment toxicity and biotic response [4]. Freezing a sample was the most effective way of preserving samples. Although the different sediment pretreatments had an influence on AVS concentrations, they had no effect on SEM levels in the sediment. AVS levels dropped during cold storage. The simplest method for preserving the original AVS conditions in sediment is to employ Ekman grab in a sediment sample and immediately store it frozen in a jar with no headspace [4, 12].

G. M. El Zokm, *Ecological Quality Status of Marine Environment*, Earth and Environmental Sciences Library, https://doi.org/10.1007/978-3-031-29203-3_3

3.2 Methodology

3.2.1 Purge and Trap Method

For evaluating SEM and AVS, the purge and trap is the most often used methodology [1, 5, 8, 13]. Because they are delicate, the purge and trap approach, which consists of complicated glass collections, is insufficient for field extractions. In this method, a round-bottom flask was joined to a trapping jar containing 0.5N NaOH (50 mL) solution (Fig. 3.1). The moist sediment sample (1 g) was added after 10 min of nitrogen gas exposure and exposed for an additional 10 min. Allen et al. [1], acidified the wet sediment sample with 20 mL 6 NHCl and agitated it at room temperature for 15 min to get AVS, gathered in a NaOH solution with a continuous nitrogen flow. A spectrophotometer was used to estimate AVS in the NaOH solution. With the evolution of AVS, the heavy metal concentrations (SEM) were determined in the filtered sample by AAS/flame [1].

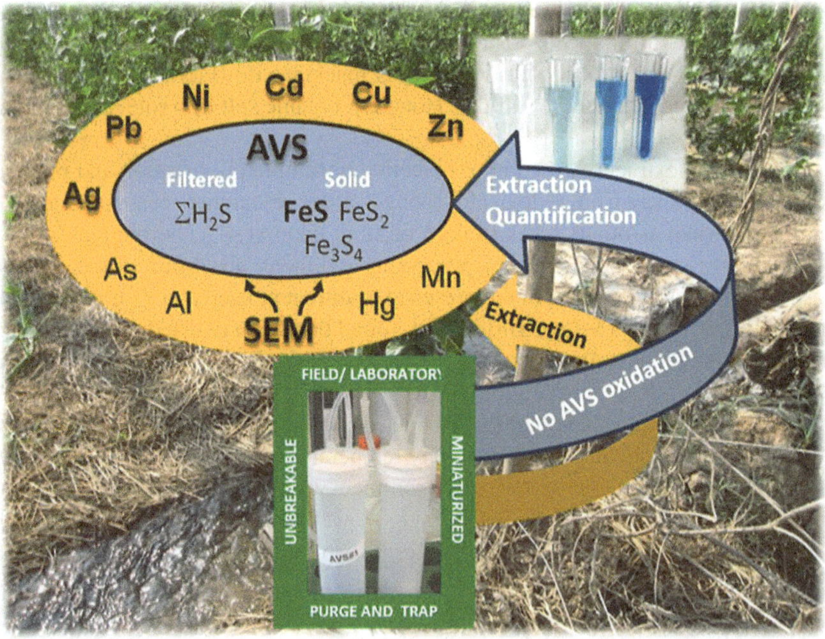

Fig. 3.1 Diagrammatic representation of field laboratory purge and trap technique for AVS extraction and simultaneously extracted metals bold divalent metals are predominant [5]

3.2.1.1 Miniaturized, Strong, and Transportable Purge and Trap

One of the most reliable ways for extraction SEM/AVS parameters on-site with a fresh sample immediately after sampling preserve the redox state. Tisserand et al. [5], proposed a new 'purge and trap' approach in the field that is small, durable, and portable (Figs. 3.1 and 3.2).

System Design and Set-Up

Optimizations and Operational Procedures

There were three steps to an extraction cycle (i) Phase purging, (ii) extraction for sulfide models component, and (iii) quantification.

(i) ***Phase of purging***

By pressurizing the system and comparing the intake and output flow rates, the airtightness of the system was determined. After that, an amount of ultrapure water

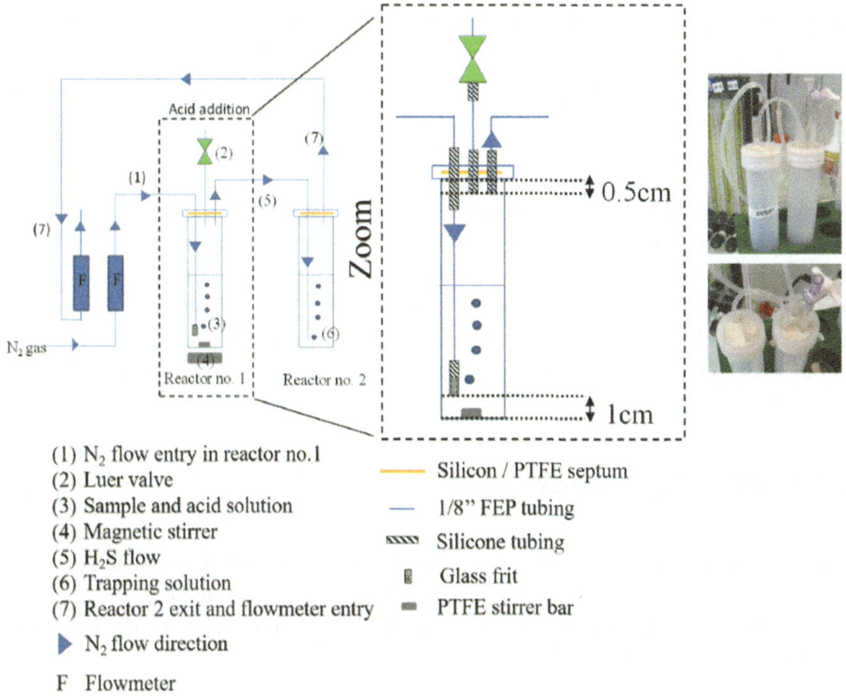

Fig. 3.2 The AVS system's 'purge and trap' plan and images. Upstream and downstream flowmeters are installed to assess the system's tightness, which is mostly comprised of Teflon® [5]

(VH_2O) was added to the mix (R1), followed by 0.5 M NaOH (VNaOH). Both reactors were closed and subjected to a flow of nitrogen gas for 5 min after Reactor No. 1 (R1) was agitated. Once the timer had expired. The gas was turned off, as well as the stirring. The cap was closed after immediately injecting and weighing the sulfide solution (V Sulfide) into R1. After that, the system was agitated for 10 min with N_2 flow (Fig. 3.2).

(ii) ***Phase of extraction***

After the purge phase was completed, HCl (20 mL) (6 or 9 M) was injected into Reactor no 1 (R1) through the Luer valve with a syringe made of polypropylene. When the Luer valve was closed in order to keep oxygen out of the reactor, the syringe was simply attached and unplugged. When injecting HCl, the flow rate of the output was constantly kept lower than the flow rate of the extracted N2 (fN2). Experiments were carried out in order to find the best flow rates and extraction times.

(iii) ***Estimations of AVS and SEM***

R2 was opened to prepare the methylene blue, 3.75 mL of MDR solution was added to react with H_2S trapped in NaOH, and reactor no 2 was manually agitated before being closed with parafilm® [14]. The MB solution's absorbance was detected at 670 nm by spectrophotometer. After extraction, the acidic solution of R1 was filtered through 0.45 mm of membrane fitter paper and measured in ICP-AES or atomic absorption. All quality assurance tests were carried out with sulfide standards and different reagent volumes (percentage of water volume to V NaOH), and HCl with 6 and 9 M.

Notes

The MDR solution was used as a trapper of H_2S in NaOH to produce MB, which was made by mixing 1 L of solution 1 with 200 mL of a solution 2; solution1 was made by combining 660 mL of 95–98% H_2SO_4 with 440 mL of water (ultra-pure), 2.25 g N–N-dimethyl-1,4-phenylen-diammonium dichloride was added once the mixture had cooled. Dilute 100 mL 37% HCl with 200 mL water (ultra-pure) and add 5.4 g of $FeCl_3.6H_2O$; Ferric chloride with 6 water molecules to make solution 2.

3.2.2 Diffusion Method

The "diffusion" approach is based on in an enclosed glass or polyethylene container, the presence of an acidified sample for a few hours to let H_2S diffuse. Diffused H_2S is subsequently contained in a solution inside the container, and S(–II) is detected using spectrophotometry in most applications.

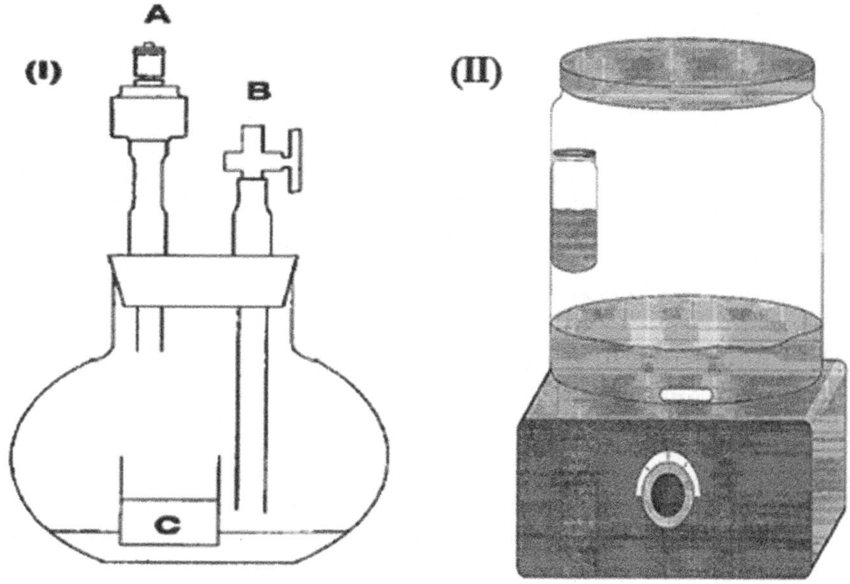

Fig. 3.3 A diagrammatic presentation of diffusion bottles of AVS system (I) solution. A = shut-off tube insert; B = syringe stopcock; C = Zinc acetate [16], (II) the flask has a 30-ml glass vial (SAOB) attached to the interior wall and a magnetic stirring [17]

3.2.2.1 Apparatus

Brouwer and Murphy [15] utilized a closed system (I) (Fig. 3.3) to acidify (1 g) soil samples and collect AVS in a sulfide antioxidant buffer (SAOB). The SAOB was tested for sulfide using an ion-selective sulfide electrode (ISE). Hsieh and Shieh [16], on the other hand, employed the equipment (I) described in (Fig. 3.3). Iodometric titration is used to measure the evolved hydrogen sulfide from 10 g of acidified wet sediment in a closed container with zinc acetate.

3.2.2.2 Reagents

The sulfide antioxidant buffer reagent (SAOB) contains 2 M NaOH to convert H_2S into S^{2-}, 1 molar ascorbic acid to inhibit S^{2-} oxidation, and 1 molar EDTA to complex metals that may have accelerated S^{2-} oxidation. Sulfide stock solution (1 M) was made from freshly cleaned sodium sulfide and kept at 4 °C. Weekly iodometric titrations against $Na_2S_2O_3$ were used. Dilutions in SAOB solution were produced daily for ISE calibration [17, 18].

3.2.2.3 Procedure

In a 50-ml scintillation vial, 1 g moist sediment and 2.5 ml water are combined. A 3.0 ml vial with SAOB is put inside the reaction flask, 2 M HCl (2.0 ml) is poured into the sediment using a bottle-top dispenser, and the vial is quickly closed to avoid H_2S loss. The completed vials are put in their original place and stirred for 1 h at 150 rpm on a rotary shaker. After that, the inner vial is eliminated, the contents are carefully mixed, and the sulfide is measured using the ion selective electrode.

3.3 Comparison of the Two Extraction Methods (Diffusion; Purge-and-Trap)

Van Griethuysen [6], designed an experiment to measure SEM and AVS in samples collected from the lakes of Ketelmeer, Nieuwersluis, and Kromme Rijn (river) in the Netherlands. Samples were used to compare the two techniques, as shown in Table 3.2. The purge-and-trap approach data (1) came from a previous study by Van den [13]. Minor differences in SEM levels are reported, indicating that the techniques are comparable in terms of extraction strength of the metal analysis. Compared with the purge-and-trap approach, the diffusion technique produces greater AVS concentrations. This was due to the diffusion method having a higher efficiency for AVS extraction compared with the purge-and-trap approach. Another possibility is that the samples' prolonged anoxic storage generated a rise in AVS [6, 19].

3.4 A New Paper Sensor Technology for Acid Volatile Sulfide Field Analysis

A method proposed by [28], reflects an improvement of the field test reported by [18] for AVS analysis in anaerobic soils. However, the new proposed procedure, is made more sensitive, accurate and much more rapid and by the use of reference color charts. The technique offers sensitivity levels comparable to laboratory techniques and is very effective for field screening.

3.4.1 Procedure of the Sulfide Paper Sensor Method

Whatman1 N.1 filter paper strips that have been treated with 6 drops of 1.5 M $Pb(NO_3)_2$ before usage and keeping 4 cm of the paper dry was used to pinch the paper strip to the jar's screw cover (Fig. 3.4). In a 250 mL polyethylene screw-cap jar, an aliquot of either soil (8 cm^3) or standardized sulfide solutions (10 mL) were

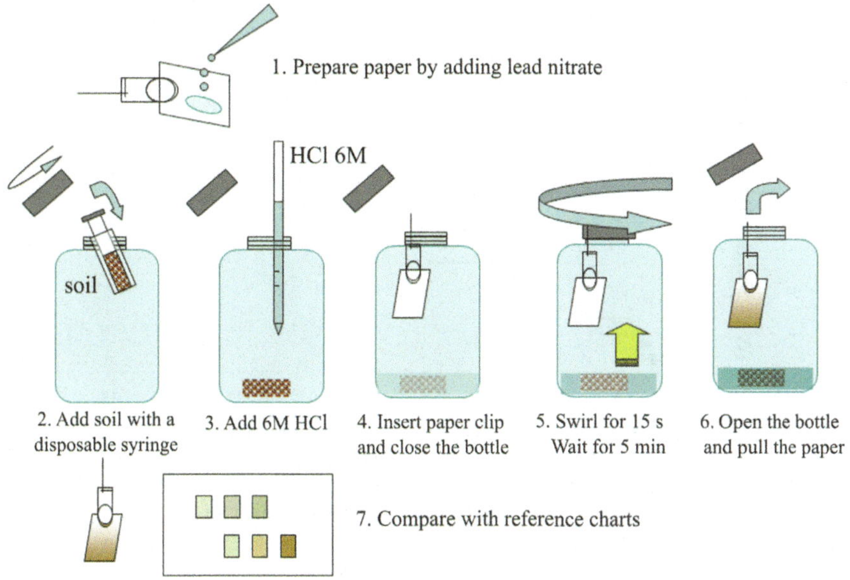

Fig. 3.4 Visual scheme of the sulfide paper sensor method for field analysis of acid volatile sulfides in soils [28]

placed. Following, the cautious addition of 50 mL of 6 M HCl along the jar's side, the cap was promptly tightened. To encourage total contact between the solution or soil and the acid and to speed up H2S volatilization, the jar was gently swirled for around 15 s. The Pb^{2+} and volatilized H_2S combine to generate PbS, which darkens the paper and gives a color that is related to the total amount of AVS and immediately compared to the reference colorimetric chart.

Since there is no recognized reference material for AVS, the soils (or sediment) used to estimate S^{2-} recovery were spiked with a known quantity of S^{2-} by adding a known volume of Na_2S. A soil sample (8 cm^3) that was anoxic and sulfide-free soil was used to test the repeatability, and 10 mL S^{2-} solutions with known concentrations were added to the soil sample. Five separate primary standard solutions, ranging in concentration from 0.2 to 3 mmol/L, were used for the test, which was carried out in triplicate on two distinct days.

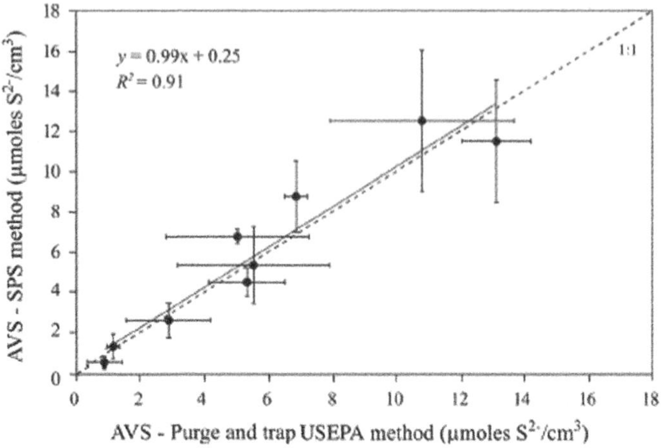

Fig. 3.5 A comparison using linear regression between the purge-and-trap method and the sulfide paper sensor method for measuring soil acid volatile sulfides concentrations [28]

3.4.2 *Sulfide Paper Sensor and Purge-and-Trap Technique Comparison*

A statistical comparison of the sulfide paper sensor method's (SPS) results with those from the purge and trap method was done (Fig. 3.5). With zero intercept, the comparison produced a very significant one-to-one linear correspondence ($R^2 = 0.99$, $p = 0.05$).

3.5 Different Methods of Quantifications of S(–II)

A lot of techniques are used for the determination of sulfide ion (Table 3.1). In most cases, S(–II) is estimated spectrophotometry [1, 5, 8, 23, 25–27]. The titration concept also is applied [16, 29]. Also, the gravimetrical technique is used [21, 22, 30] or measured directly with a sulfide ion-selective electrode [6, 17, 24, 29;4] (Table 3.2).

Table 3.1 Extraction techniques and experimental settings for AVS model

References	Method	[HCl] (M)	Cold (C)/hot (H) HCl	Extraction	Trapping solution	S(–II) quantification
[20]	Purge and trap	1	C	30 min	ZnAc/acetic acid	Methylene blue
[1]	Purge and Trap	1	C	40 min	NaOH	Methylene blue
[21, 22]	Purge and trap	0.5	C	1 h	0.1 M AgNO3	Methylene blue
[13]	Purge and trap	6	C	> 45 min	NaOH	Colorimetric/ potentiometric
[23]	Purge and trap	1	C	> 1 h	ZnAc/NaAc	Methylene blue
[24]	Purge and trap	1	C	1–2 h	NaOH	Ion probe
[6]	Diffusion	1	C	4 h	NaOH/ascorbic acid/EDTA	Ion probe
[6]	Diffusion	1	C	4 h	NaOH/ascorbic acid/EDTA	Ion probe
[25]	Purge and trap	6	C & H	1 h	ZnAc	Methylene blue
[26]	Diffusion	1	C	4 h	NaOH	Methylene blue
[4]	Diffusion	1	C	4 h	NaOH/ascorbic acid/EDTA	Ion probe
[27]	Purge and trap	12	C	1 h	ZnAc/NH_4OH	Methylene blue
[8]	Purge and trap	6	C	15 min	NaOH	Methylene blue
[5]	Purge and trap	6 or 9	C	(30, 60 and 90 min)	NaOH	Methylene blue

ZnAc, NaAc stand for zinc acetate and sodium acetate, $AgNO_3$ stands for silver nitrate, and EDTA stands for ethylene diamine tetra acetic acid [5] with permission

Table 3.2 A comparison of the purge-and-trap approach (1) with the diffusion method (2) for determining AVS and SEM concentrations [6] with permission

Sample	Method			
	Purge and trap		Diffusion	
	AVS	∑SEM	AVS	∑SEM
Ketelmeer (N = 5)	10.0	10.5	26.8 ± 5.1[a]	11.3 ± 1.3
Nieuwersluis (N = 5)	19.8	2.9	27.8 ± 8.1	3.5 ± 0.9
Kromme Rijn (N = 5)	21.7	9.6	27.9 ± 1.6[a]	9.9 ± 0.7

For the diffusion method, 1–2 h is the reaction time of for the Nieuwersluis sample and 4 h for Kromme and the Ketelmeer Rijn samples. The number of replicate measurements (N = 5) refers to the diffusion method.

[a] Value significant difference ($\alpha < 0.05$) from the value obtained from purge-and-trap

References

1. Allen HE, Fu G, Deng B (1993) Analysis of acid-volatile sulfide (AVS) and simultaneously extracted metals (SEM) for the estimation of potential toxicity in aquatic sediments. Environ Toxicol Chem 12:1441–1453. https://doi.org/10.1002/etc.5620120812
2. Meysman JRP, Middelburg JJ (2005) cid-volatile sulfide (AVS)—a comment. Mar Chem 97:206–212. https://doi.org/10.1016/j.marchem.2005.08.005
3. Rickard D, Morse JW (2005) Acid volatile sulfide (AVS). Mar Chem 97(3–4):141–197. https://doi.org/10.1016/j.marchem.2005.08.004
4. De Lange HJ, van Griethuysen C, Koelmans AA (2008) Sampling method, storage and pretreatment of sediment affect AVS concentrations with consequences for bioassay responses. Environ Pollut 151:243–251. https://doi.org/10.1016/j.envpol.2007.01.052
5. Tisserand D, Guédron S, Razimbaud S, Findling N, Charlet L (2021) Acid volatile sulfides and simultaneously extracted metals: a new miniaturized 'purge and trap' system for laboratory and field measurements. Talanta 233:22490. https://doi.org/10.1016/j.talanta.2021.122490
6. Van Griethuysen C, Gillissen F, Koelmans A (2002) Measuring acid volatile sulphide in floodplain lake sediments: effect of reaction time, sample size and aeration. Chemosphere 47(4):395–400. https://doi.org/10.1016/s0045-6535(01)00314
7. He J, Lü C, Fan Q, Xue H, Bao J (2011) Distribution of AVS-SEM, transformation mechanism and risk assessment of heavy metals in the Nanhai Lake in China. Environ Earth Sci 64(8):2025–2037. https://doi.org/10.1007/s12665-011-1022-z
8. El Zokm GM, Okbah MA, Younis AM (2015) Assessment of heavy metals pollution using AVS-SEM and fractionation techniques in Edku Lagoon sediments, Mediterranean Sea. J Environ Sci Health Part A Toxic/Hazard Subst Environ Eng 50:571–584. https://doi.org/10.1080/10934529.2015.994945
9. Lasorsa B, Casas A (1996) A comparison of sample handling and analytical methods for determination of acid volatile sulfides in sediment. Mar Chem 52:211–220. https://doi.org/10.1016/0304-4203(95)00074-7
10. Billon G (2001) Chemical speciation of sulfur compounds in surface sediments from three bays (Fresnaye, Seine and Authie) in northern France, and identification of some factors controlling their generation. Talanta 53(5):971–981. https://doi.org/10.1016/s0039-9140(00)00586
11. DeFoe DL, Ankley GT (1998) Influence of storage time on toxicity of freshwater sediments to benthic macroinvertebrates. Environ Pollut 99:123–131. https://doi.org/10.1016/S0269-7491(97)00159-0
12. Elhaj Baddar Z, Peck E, Xu X (2021). Temporal deposition of copper and zinc in the sediments of metal removal constructed wetlands. Publ Libr Scie (PLoS) One 16(8):0255527. https://doi.org/10.1371/journal.pone.0255527
13. Van den Hoop MAGT, Den Hollander HA, Kerdijk HN (1997) Spatial and seasonal variations of acid volatile sulfide (AVS) and simultaneously extracted metals (SEM) in Dutch marine and freshwater sediments. Chemosphere 35(10):2307–2316. https://doi.org/10.1016/S0045-6535(97)00309-3
14. Reese BK, Finneran DW, Mills HJ, Zhu MX, Morse JW (2011) Examination and refinement of the determination of aqueous hydrogen sulfide by the methylene blue method. Aquat Geochem 17:567–582. https://doi.org/10.1007/s10498-011-9128-1
15. Brouwer H, Murphy TP (1994) Diffusion method for the determination of acid-volatile sulfides (AVS) in sediment. Environ Toxicol Chem 13(8):1273–1275. https://doi.org/10.1002/etc.5620130808
16. Hsieh YP, Shieh YN (1997) Analysis of reduced inorganic sulfur by diffusion methods; improved apparatus and evaluation for sulfur isotopic studies. Chem Geol 137:255–261. https://doi.org/10.1016/S0009-2541(96)00159-3
17. Leonard EN, Cotter AM, Ankley GT (1996) Modified diffusion method for analysis of acid volatile sulfides and simultaneously extracted metals in freshwater sediment. Environ Toxicol Chem 15(9):1479–1481. https://doi.org/10.1002/etc.5620150908

18. Anderson EF, Wilson DJ (2000) A simple field test for acid volatile sulfide in sediments. J Tennessee Acad Sci 74(3–4):53–56. PMID: 11764145
19. Van Griethuysen C, Meijboom EW, Koelmans AA (2003) Spatial variation of metals and acid volatile sulfide (AVS) in floodplain lake sediment. Environ Toxicol Chem 22(3):457–465. https://doi.org/10.1002/etc.5620220301
20. Fossing H, Jørgensen BB (1989) Measurement of bacterial sulfate reduction in sediments: evaluation of a single-step chromium reduction method. Biogeochemistry 8:205–222. https://doi.org/10.1007/BF00002889
21. Di Toro DM, Mahony JD, Hansen DJ, Scott KJ, Hicks MB, Mayr SM, Redmond MS (1990) Toxicity of cadmium in sediments: the role of acid volatile sulfide. Environ Toxicol Chem 9(12):1487–1502. https://doi.org/10.1002/etc.5620091208
22. Berry WJ, Hansen DJ, Mahony JD, Robson DL, Di Toro DM, Shipley BP, Rogers B, Corbin JM, Boothman WS (1996) Predicting the toxicity of metal-spiked laboratory sediments using acid-volatile sulfide and interstitial water normalizations. Environ Toxicol Chem 15(12):2067–2079. https://doi.org/10.1002/etc.5620151203
23. Huerta-Diaz MA, Tessier A, Carignan R (1998) Geochemistry of trace metals associated with reduced sulfur in freshwater sediments. J Appl Geochem 13:213–233. https://doi.org/10.1016/S0883-2927(97)00060-7
24. Billon G, Ouddane B, Boughriet A (2001) Artefacts in the speciation of sulfides in anoxic sediments. Analyst 126:1805–1809.https://doi.org/10.1039/b104704n
25. Praharaj T, Fortin D (2004) Determination of acid volatile sulfides and chromium reducible sulfides in Cu–Zn and Au mine tailings. Water Air Soil Pollut 155:35–50. https://doi.org/10.1023/B:WATE.0000026526.26339.c3
26. Zhang X, Zhang L (2007) Acid volatile sulfide and simultaneously extracted metals in tidal flat sediments of Jiaozhou Bay, China. J Ocean Univ China 6:137–142. https://doi.org/10.1007/s11802-007-0137-z
27. Rees GN, Baldwin DS, Watson GO, Hall WK (2010) Sulfide formation in freshwater sediments, by sulfate-reducing microorganisms with diverse tolerance to salt. Sci Total Environ 409:134–139. https://doi.org/10.1016/j.scitotenv.2010.08.062
28. Pellegrini E, Contin M, Vittori Antisari L, Vianello G, Ferronato C, De Nobili M (2018) A new paper sensor method for field analysis of acid volatile sulfides (AVS) in soils. Environ Toxicol Chem 37(12):3025–3031. https://doi.org/10.1002/etc.4279
29. Gagnon C, Mucci A, Pelletier E (1995) Anomalous accumulation of acid-volatile sulphides (AVS) in a coastal marine sediment, Saguenay Fjord, Canada. Geochimica et Cosmochimica Acta 59(13):2663–2675. https://doi.org/10.1016/0016-7037(95)00163-T
30. Di Toro DM, Mahony JD, Hansen DJ, Scott KJ, Carlson A, Ankley GT (1992) Acid volatile sulphide predicts the acute toxicity of cadmium and nickel in sediments. Environ Sci Technol 26:96–101. https://doi.org/10.1021/es00025a009

Chapter 4
AVS-SEM Models

The interaction of metal contaminants with AVS, which results in the formation of metal sulfides, is a crucial process that regulates the bioavailability of many metals in sediments [1–5]. The AVS/SEM models predict that metals will not produce direct toxicity to benthos if AVS exceeded SEM in molar concept. Under geochemical conditions that restrict the free metal level, metal bioavailability is a function of metal/particle interaction and organism activity [6]. On an AVS basis, the equilibrium partitioning theory (EqP) can be used to connect cationic metal concentrations in sediment to a shortage of readily dissolved amounts in interstitial water [7].

4.1 The Mechanism Controlling AVS-SEM Systems

Manganese mono sulfides (MnS) and Iron mono sulfides (FeS) have vital action in determining the metal bioavailability in anaerobic sediments. MnS and FeS represent the most soluble minerals and have been methodology defined as AVS [8, 9]. There are two paths for Sulfide production.

a. Direct bacterial reduction.
b. Indirect bacterial action of organic matter as acceptor of electron

As FeS and MnS have higher Ksp than any other heavy metal sulfide, metals are going to replace Fe and Mn to produce more insoluble MeS (Table 4.1).

$$Me^{2+} + FeS_{(s)} \rightarrow MeS_{(s)} + Fe^{2+} \tag{4.1}$$

$$Me^{2+} + MnS_{(s)} \rightarrow MeS_{(s)} + Mn^{2+} \tag{4.2}$$

The formed MeS can be either pure MeS or co-precipitated or adsorption on FeS [8–10]. If all metals in sediments are in the MeS form, the metal ion's mobility in pore fluids will be controlled by MeS dissolution. Environmental metals' mobility will be

G. M. El Zokm, *Ecological Quality Status of Marine Environment*, Earth and Environmental Sciences Library, https://doi.org/10.1007/978-3-031-29203-3_4

Table 4.1 Solubility products of metal sulfide related to FeS and MnS [11] with permission

Metal sulfide	K_{sp}	$K_{sp\,(MeS)}/K_{sp\,(FeS)}$	$K_{sp\,(MeS)}/K_{sp\,(MnS)}$
FeS	2.29×10^{-4}	–	5.75×10^{-4}
MnS	3.98×10^{-1}	1.74×10^{3}	–
PbS	9.04×10^{-29}	3.9×10^{-25}	2.27×10^{-28}
CuS	6.31×10^{-23}	2.76×10^{-19}	1.59×10^{-22}
CdS	7.94×10^{-15}	3.47×10^{-11}	2×10^{-14}
NiS	5.89×10^{-10}	2.57×10^{-6}	1.48×10^{-9}
ZnS	1.58×10^{-11}	6.9×10^{-8}	3.97×10^{-11}
CoS	3.98×10^{-8}	1.74×10^{-4}	1×10^{-7}

restricted in the presence of abundant AVS because their sulfide solubility products are lower than those of FeS. As a result, the metals are unlikely to be harmful if there is an excess of AVS.

4.2 Acid Volatile Sulfide as a Heavy Metal Toxicity Controller in Marine Sediments

The reduction of iron oxides, produces iron sulfides as a precipitating agent for heavy metals. Sulfate-reducing bacteria generate hydrogen sulfide. The iron sulfide minerals are divided into two categories: (a) AVS; the solid-phase sulfides inclusive amorphous (FeS), mackinawite (FeS), and greigite (Fe_3S_4) and (b) pyrite (FeS_2) [12, 13]

$$\mathbf{2}\,CH_20 + SO_4{}^{2-} \rightarrow H_2S + 2HCO_3{}^- \tag{4.3}$$

$$2FeOOH + 3H_2S \rightarrow 2FeS + S^0 + 4H_2O \tag{4.4}$$

$$FeS_{amorphous} \rightarrow FeS_{mackinawite} \tag{4.5}$$

$$3FeS_{mackmawite} + S^{\circ} \rightarrow Fe_3S_4\ \text{greigite} \tag{4.6}$$

$$Fe_3S_{4\,greigite} + 2\,S^0 \rightarrow 3FeS_{2\,pyrite} \tag{4.7}$$

Because transformation rates are modest, metastable species can remain in sediment for long periods (AVS). The crucial stage involves the precipitation of metal sulfides. Many researchers [14, 15] have looked at the interactions between heavy metals and AVS to figure out how intractable metal sulfides are formed. Metal sulfides that were less soluble than FeS began to develop when metal concentrations rose in sediment.

The solubility of metal cations increases in the following order

$$Co > Ni > Zn > Cd > Cu$$

Because of the decrease of pore water sulfides, solid-phase iron sulfides dissolved and were bound to other metals were thought to be the source of free sulfide.

$$Me^{2+} + FeS(s) \rightarrow MeS(s) + Fe^{2+} \tag{4.8}$$

$$Ni^{2+} + CoS(s) \rightarrow NiS(s) + Co^{2+} \tag{4.9}$$

$$Zn^{2+} + Ni\,S(s) \rightarrow Zn\,S(s) + Ni^{2+} \tag{4.10}$$

$$Cd^{2+} + ZnS(s) \rightarrow Cd\,S(s) + Zn^{2+} \tag{4.11}$$

$$Cu^{2+} + CdS(s) \rightarrow Cu\,S(s) + Cd^{2+} \tag{4.12}$$

Because the formation of an insoluble metal species significantly reduces metal bioavailability, no toxic effect will be found in tested organisms as measured by mortality [16]. Metal sulfides have lower solubility than their hydroxide and carbonate equivalents, allowing for more thorough precipitation and greater stability over a wide pH range. Sulfide precipitation offers several benefits over hydroxide and carbonate precipitation: 1—There is less chelation agent interference in waste water; 2—There is better precipitation selectivity; 3—There are high reaction rates; and 4—There is less sulfide improved thickening and drying abilities of their sludge rather than hydroxide sludge [17, 18].

4.3 Profiles AVS and SEM with Depths (Core Sediment)

Sediment core analysis has proven to be a highly effective method for determining the impact of SEM and AVS on depositional settings over the past few decades. Many researchers studied AVS and SEM variations as a function of depths in core sediment [19–22; 42; 71]. Diao et al. [19], reported special apportionment of AVS and SEM in in Lake Chaohu, China with Fig. 4.1 a low level of AVS in sediments at depths of 1 cm below the Sediment water interface, after that the AVS increased with the sediment profile depths. however, the corresponding ΣSEM levels were relatively constant compared to those of AVS this distribution is in agree with van den Berg et al. (1998), [42]. The dynamic equilibrium between AVS formation and loss by oxidation or diffusion leads to the variation of AVS in core sediment, which is vulnerable to a number of processes, including bioturbation, sediment resuspension, rate of sulphate reduction, sediment oxidation–reduction, and organic matter level [23]. The upper sediment near had a low concentration of AVS with a high oxygen content due to the frequent bioturbation, resuspension, and re-sedimentation processes, whereas the high concentration of AVS in the deeper sediment could be explained by the restoration of sulfate in an anaerobic environment [19]. AVS, SEM, and TOC concentrations were found to be strongly significantly correlated. Wang et al. [21], displays the AVS and SEM depth profiles in sediment cores from the four sites. SEM was evidently more equally distributed than AVS because the fluctuations in SEM concentrations in sediment cores were substantially lower than those for AVS. every sediment. AVS concentration varied greatly throughout the depth of cores, being low in the first few centimeters (0–2 cm), increasing suddenly to reach a maximum at 10 cm, and then gradually declining in deeper sediments as the core profile descended Fig. 4.2. These findings were consistent with recent research showing a fluctuation in AVS concentration with depth.

A study performed by [22], in revealed that the AVS peak values in the sediment profile shifted closer to the sediment's surface with increased organic matter contents and deeper water depth (Fig. 4.3).

Because of the high sulfate content in the water bodies' potential migration to the deeper segment of the sediment profile, the AVS peak values shift deeper in the sediment profile as sulfate concentration increases.

Machado et al. [24], studied the vertical distribution of three sulfide models in two revers; São João de Merití River and Iguaçu River (Fig. 4.4). However, [25] examined the Seasonal and vertical apportionment of AVS/SEM relationship in sediments and potential bioavailability of metals in estuary, southeastern Brazil [Morrão and Cubatão Riverw] (Fig. 4.5).

Another approach correlates the total metals to SEM in core sediment in the Pearl River Estuary, South China was examined by [26]. The distribution of SEM and total metals in sediment core is presented in Fig. 4.6. The SEM concentrations can be used to evaluate any potential relationships between heavy metals and sulfides in the sediment cores. SEM's vertical apportionment and the total metal concentrations in Core sediment followed the same trend, demonstrating a good correlation between the SEM and the total metals in the sediments.

4.4 Evaluation of Sediment Toxicity

By assessing the "bioavailable" proportion of metals present, the AVS/SEM method gives a more precise estimation of sediment toxicity to benthic organisms (than criteria for "total" metals). The sum of the SEM (μmol/g dry weight) and the AVS concentration is used to determine the toxicity of the sediment. If $\sum$SEM have lower concentrations than AVS, no risk; if $\sum$SEM have greater concentrations than AVS, divalent metals may exist as free metal ions and cause toxicity [8, 28–30; 27]. The biotic ligand model (BLM), equilibrium partitioning model (EqP), and the

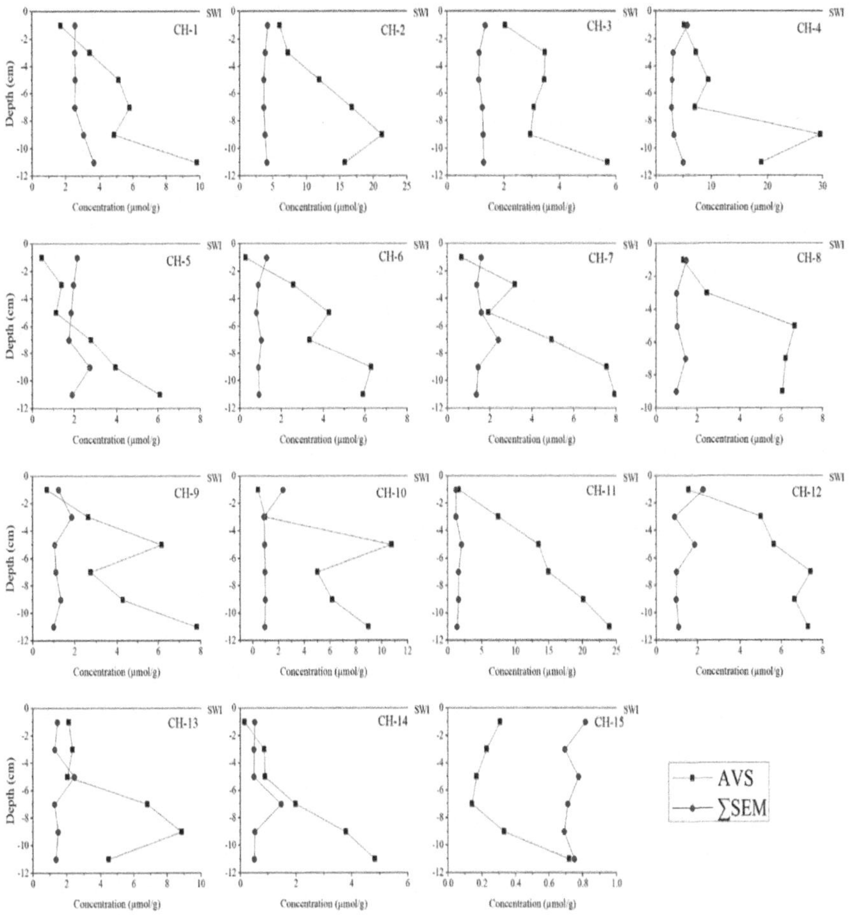

Fig. 4.1 Vertical distributions of concentrations of AVS and ΣSEM in the sediments from Lake Chaohu [19] with permission

chemistry of metal–sulfide interactions are used in SEM-AVS models to predict toxicity [3, 27, 28].

4.4.1 Assessment of Toxicity Using SEM-AVS Models

The AVS/SEM ratio has been recommended by the USEPA as a predictor of metals' bioavailability in sediment. These models recognize that many toxic heavy metals form insoluble sulfides in sediment, reducing metal bioavailability. When metal molar concentrations exceed AVS (sulfide buffering capacity), heavy metals are released into the interstitial water and can cause toxicity. SEM-AVS models can be used to

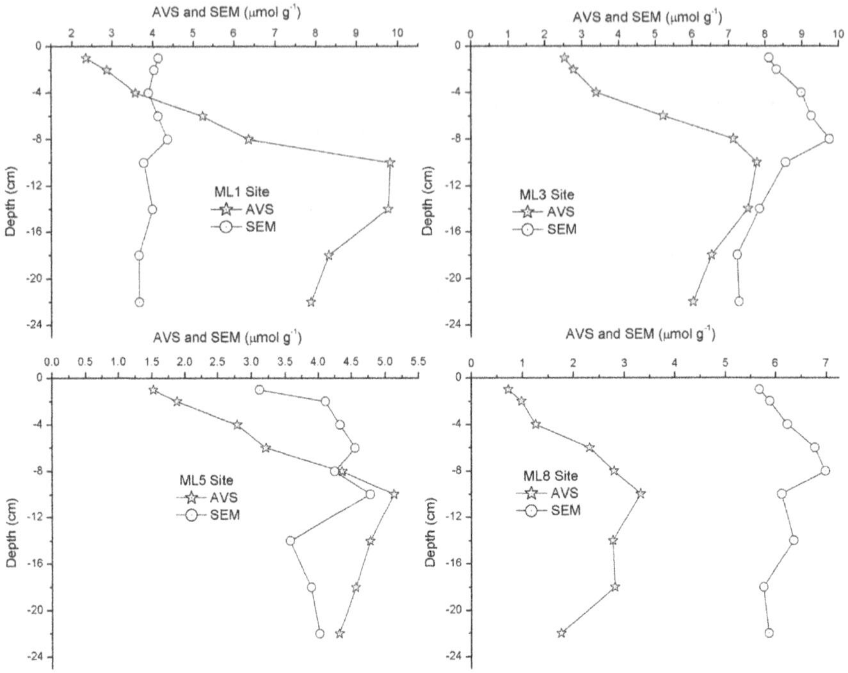

Fig. 4.2 Profile of AVS and SEM in the sediment cores from different sites of Maluan Bay [21] with permission

manage metal-contaminated sediments by simplifying and facilitating the assessment of metal toxicity in field circumstances.

Except for silver, where the ratio is 2:1, the stoichiometry of AVS absorption of divalent metals is such that 1 mol of AVS will stabilize 1 mol of SEM, allowing the use of AVS models to forecast the absence of toxicity. In reality, regardless of the specifics of the sediment chemistry, it is the very poor solubility of the resultant metal sulfides that keeps interstitial water concentrations below hazardous levels (e.g., pH, iron concentration). Finally, AVS can accurately assess the hazards posed by heavy metals, with additional benefits such as high efficiency, ease of use, and batch processing [29–31]. In order to evaluate heavy metal toxicity, AVS-SEM models were proposed; ΣSEM/AVS [32], ΣSEM-AVS [33], (ΣSEM-AVS)/*f*OC [3], AVS/Fe [34], $RQ_{SEM\text{-}TEL} = \sum (C_{SEM}/TEL)$ and $RQ_{SEM\text{-}PEL} = \sum (C_{SEM}/PEL)$ [35]. However, AVS value can be used for classifying the lakes as four categories; the healthy area (AVS < 6.86 μmol/g), the eutrophic area (AVS; 6.86–13.41 μmol/g), the warning area (AVS; 13.41–22.14 μmol/g) and the critical area (AVS < 22.14 μmol/g) [36].

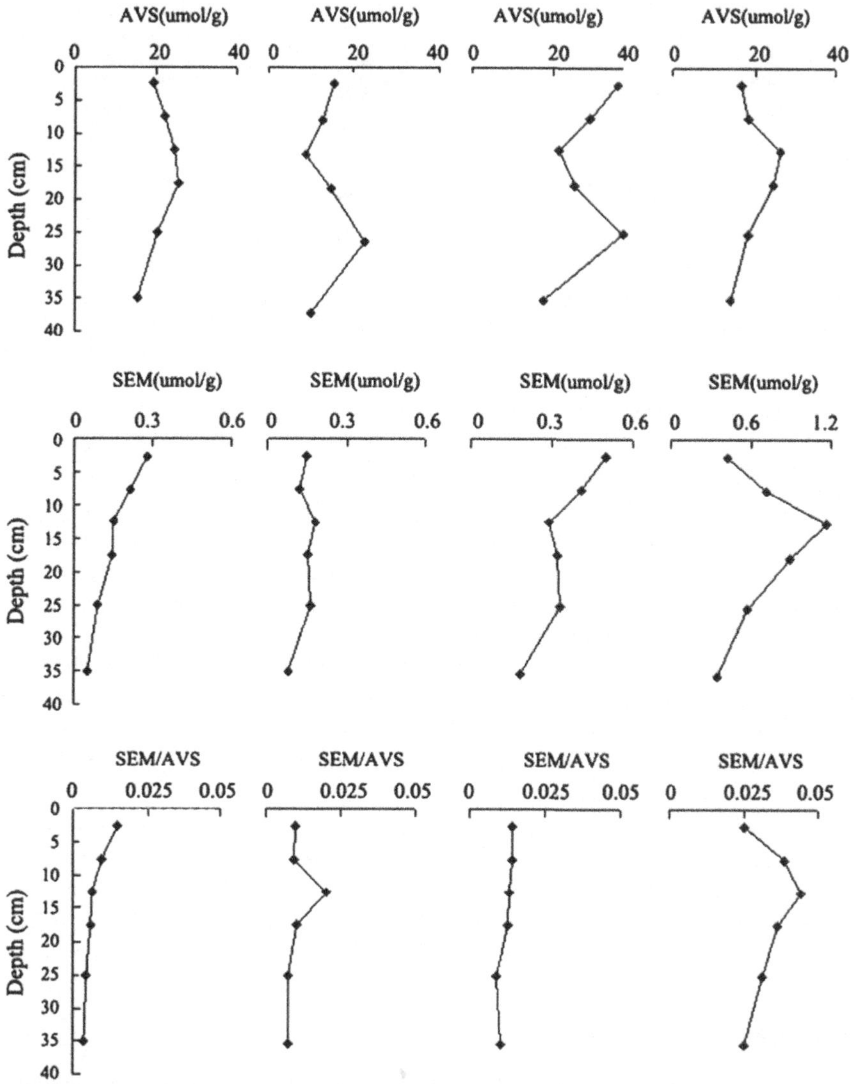

Fig. 4.3 AVS, SEM and [SEM]/[AVS] variation with depth in core sediments from Nanhai Lake in China [22]

4.4.1.1 ΣSEM/AVS Model with Reference to BLM and EqP Theories

It is assumed that dietary metal buildup is at least as significant as ion transport from the aqueous phase and that it often exceeds it. The amount of possible metal toxicity to sediment-dwelling species is indicated by an SEM/AVS ratio > 1 [14, 15]. Metals have a damaging influence on benthic macroinvertebrates and other aquatic

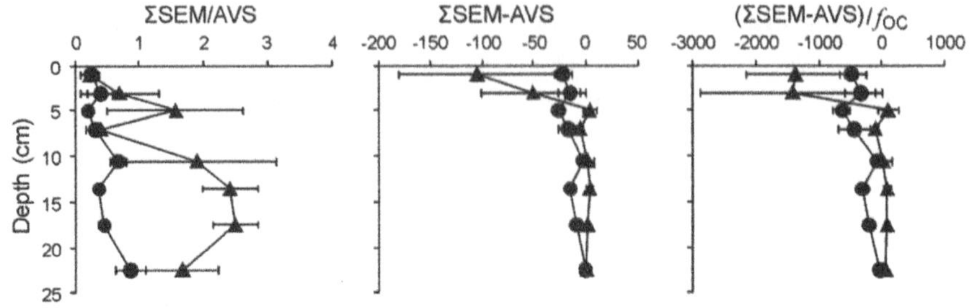

Fig. 4.4 (ΣSEM/AVS, ΣSEM-AVS and (ΣSEM-AVS)/*fOC*) in sediment cores from Iguaçu River (circles) and São João de Merití River (triangles) estuaries [24]

life when the SEM/AVS value is greater than 2 [31, 37]. The BLM as a method for forecasting metals' ecotoxicological effects in the environment has been used [38]. From a biological viewpoint, mass transfer across the plasma membrane is the most significant metal intake pathway. Figure 4.7 depicts three different exposure scenarios: free metal ions, metal complexes, and particle-bound metals. When the rate of metal detoxification and excretion is less than the rate of metal absorption from all sources, toxicity occurs. BLM stated that significant metal uptake originates from a pool of free metal ions (Fig. 4.8).

Shine et al. [40], studied the SEM/AVS model's applicability. This method has very high sensitivity, according to the SEM/AVS model evaluation findings (96%), in accurately classifying a hazardous sample as harmful and thus being environmentally protective. The capacity to anticipate both positive and negative outcomes, in addition to sensitivity, was investigated.

The [SEM-AVS] controls aquatic creature toxicity explained by equilibrium partitioning theory (EqP). Even so, many EqP-based studies did not account for tissue biomagnification and weren't able to display a meaningful link between bioaccumulation and the [SEM-AVS]. Previous Eq P-based studies linked toxicity effects to [SEM-AVS], but biomagnification data appeared to be closely linked with [SEM] [41–43].

EqP theory

The argument that AVS limits activity and, consequently, the metal bioavailability in anoxic sediments is a feature of the EqP concept. On an AVS approach, EqP theory connects metal concentrations in sediments to dissolved amounts in interstitial water. Chemical bioavailability to sediment-dwelling organisms is considered to be proportional to metal mobility in the sediment, as measured by metal ion concentration in interstitial water. Sulfide is a key partitioning fraction in anoxic sediments that controls cationic metal flux in the sediment–interstitial water system [7, 41, 44].

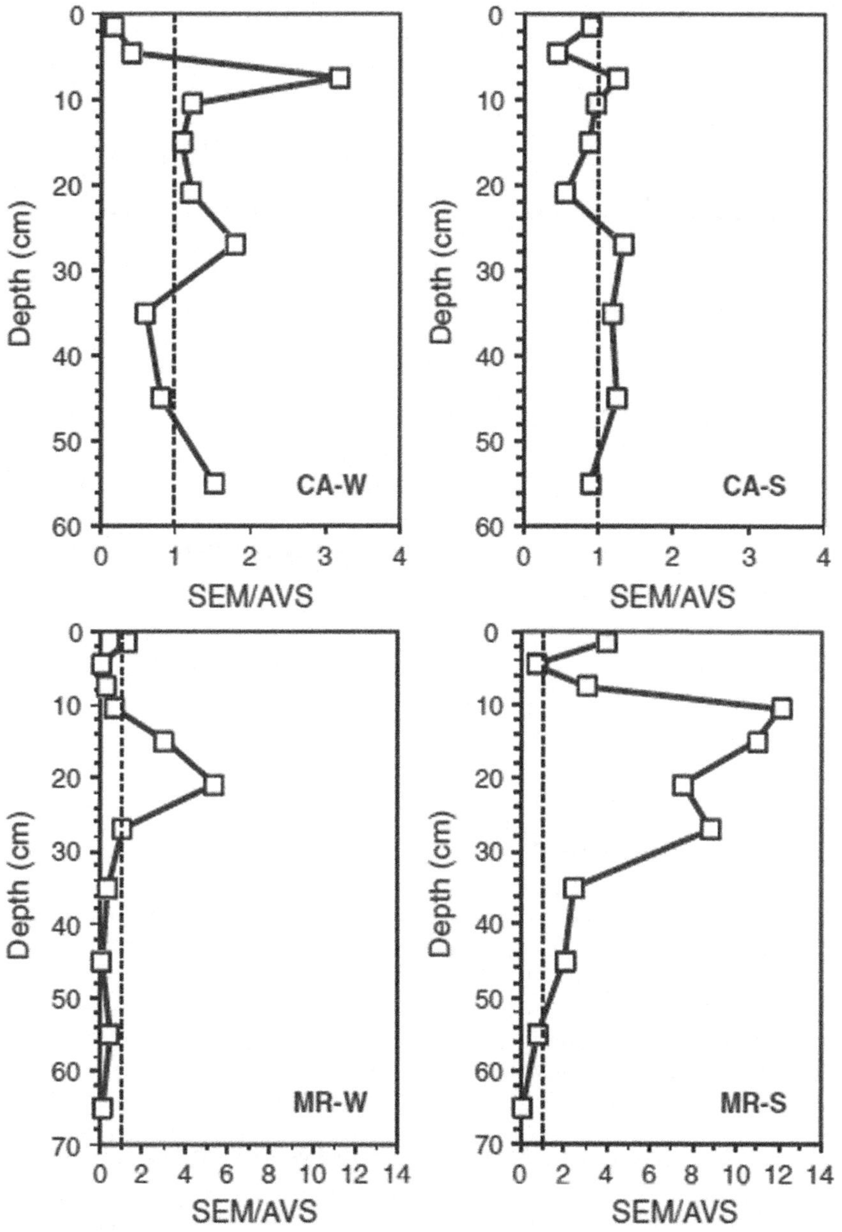

Fig. 4.5 Relationship between the SEM and AVS molar concentrations in sediment cores of the Casqueiro (CA-W = winter and CA-S = summer) and Morrão (MR-W = winter and MR-S = summer) [25]

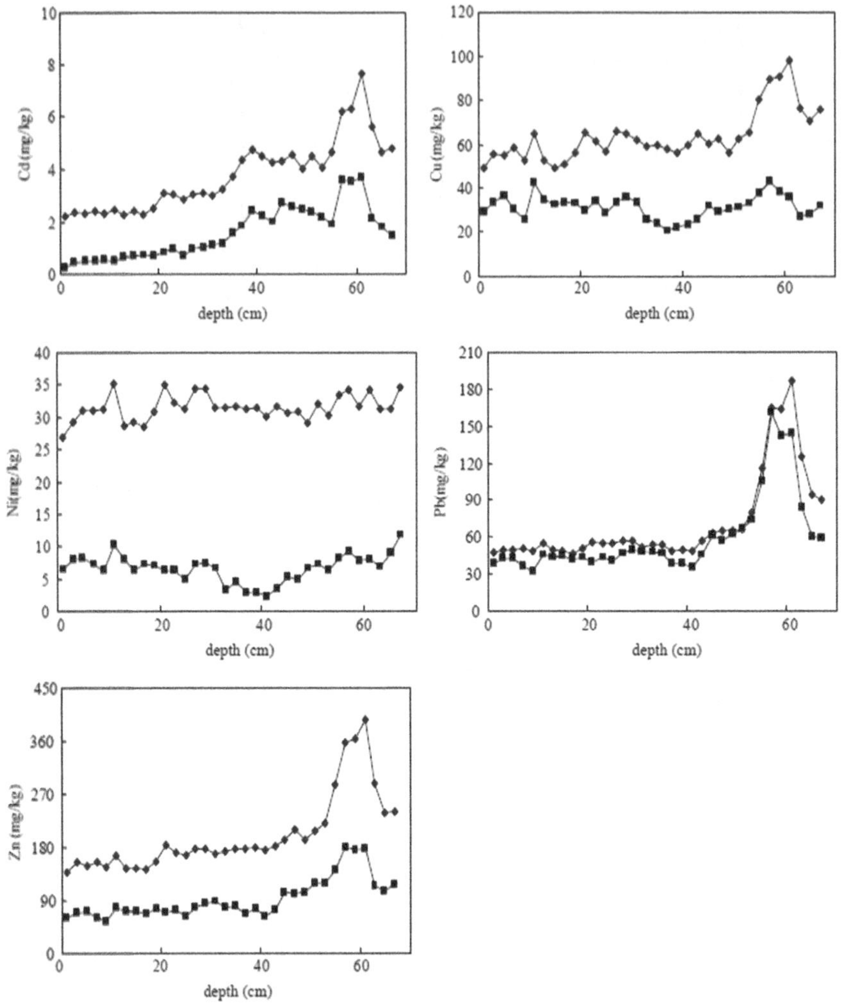

Fig. 4.6 SEM and total level of Ni, Cd, Pb, Zn and Cu in Core1 sediment of Pearl River Estuary, South China (◆ total metal; ■ SEM) [26] with permission

4.4.1.2 Derivation of ΣSEM-AVS Model Based on EqP Theory

The EqP model was introduced by Di Toro et al. (2002), as a first approach for determining the critical metal concentrations that forecast toxicity in sediments. The metal concentration in sediment C_S coincides with a LC_{50} in water when exposed to the tested organism i.

$$CS^* = Kp\,LC_{50} \tag{4.13}$$

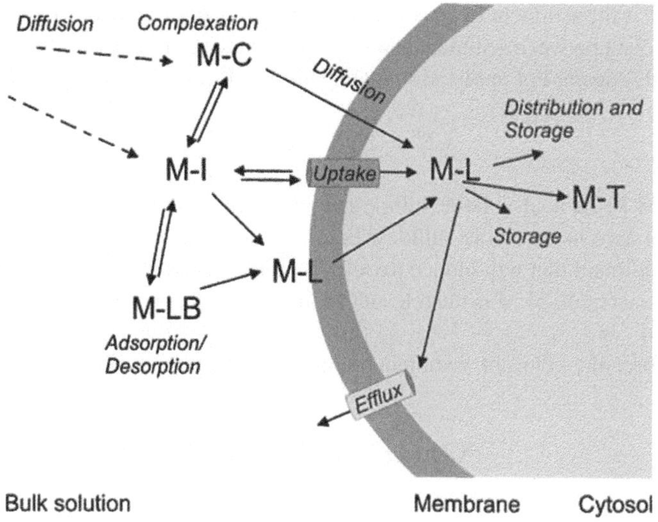

Fig. 4.7 The key mechanisms and sources for trace metal absorption at a biological membrane are depicted in this conceptual model. Metal ion (M-I), metal complex (M-C), labile particle-bound metal (M-LB), metal-bound (M-L) to a biological ligand, M-T metal at a target site, M-T metal at the target site, M-L metal bound to a biological ligand, [39]

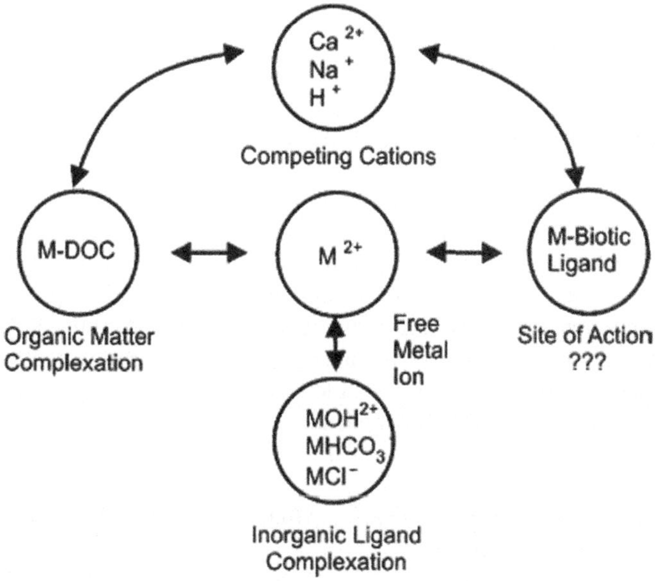

Fig. 4.8 The biotic ligand model design [39]

where CS* is the sediment LC_{50} concentration (μg/kg dry wt), Kp (L/kg) is the partition coefficient between sediment and interstitial water and LC_{50} is the concentration causing 50% death. For application metal sulfides, Eq. 4.13 becomes

$$CS^* = AVS + Kp\,LC_{50} \tag{4.14}$$

where AVS is the acid volatile sulfide content in the sediment. Because AVS may bind metal as very insoluble sulfides, Eq. 4.6 simply indicates that the quantity of metal in sediment that will induce toxicity is at least as much as the amount of AVS present. As a result, SEM is the relevant sediment metal concentration, and Eq. 4.14 is modified.

Therefore, the relevant sediment metal concentration is SEM, and Eq. 4.14 becomes

$$\sum SEM = AVS + Kp\,LC_{50} \tag{4.15}$$

The AVS hypothesis is based on the observation that if Eq. 4.15 's second part is ignored, the critical status is achieved when SEM = AVS, and the criteria for toxicity or absence of toxicity is SEM-AVS = 0.0. (μmol/g dry wt) according to the [32] evaluation. The recommended sediment quality criterion formula can be used to estimate whether partitioning to other phases should be addressed (Di Toro et al. 2002):

$$[C_{SQC}] = [AVS] + K_P[C_{WQC}] \tag{4.16}$$

In the case of a 1μ mol/g (very little) AVS concentration,

$$C_{SQC} = K_P C_{WQC} \tag{4.17}$$

Due to partitioning, a metal with a criterion concentration of 1 mol/L will have its sediment quality criteria twice. By comparing the product Kp × [CWQC] to the AVS concentration, the importance of partitioning may be determined. As a result, partitioning only has an influence on metals with higher partition coefficients and chronic water quality criterion values at low AVS concentrations (1μ mol AVS/g) and metals with higher partition coefficients.

4.4.1.3 Derivation of $\sum$SEM-AVS/f_{OC} Model Based on EqP Theory and Organic Carbon

To better toxicity prediction approximating the partitioning term rather than ignore it is recommended [3]. Partition coefficients for certain metals at various pHs have been determined for the organic carbon fraction, which is a key partitioning phase in sediments (Mahony et al. 1991). This means that Kp in Eq. 4.17 may be described using the organic carbon–water partition coefficient, K_{OC}, and the fraction of organic carbon in the sediment, f_{OC}.

$$Kp = foc \times K_{OC} \tag{4.18}$$

Using this expression in Eq. 4.18 yields

$$\sum SEM = AVS + foc K_{OC} LC_{50} \tag{4.19}$$

Moving the known terms to the left side of this equation yields

$$\left[\sum SEM - AVS\right] / foc = K_{OC} LC_{50} \tag{4.20}$$

Equation 4.20 may be used to estimate toxicity if both K_{OC} and LC_{50} are known. According to the Environmental Protection Agency (EPA), equilibrium partitioning sediment benchmarks (ESBs) are a way to gauge toxicity based on the bioavailable metal fraction, which can be evaluated in pore water and/or anticipated using relative AVS, SEM, and organic carbon (OC) in sediment. Because AVS and organic carbon may immobilize a variety of heavy metals in sediments [33], As normalized factor, OC is used to the difference between SEM and AVS which has been widely utilized to measure metal toxicity in sediment. However, not all sediments with a [SEM]/[AVS] ratio greater than 1 can be identified as toxic since, other metal-binding components found in sediments include OC and Fe/MnOx [9, 41].

As a result, environmental ratios like [Mn]/[AVS] and [Fe]/[AVS] should be considered. This might be why [SEM]/[AVS] ratios overestimate metal availability [41, 45, 46]. When [SEM]/[AVS] > 1, alternative metal binding forms in sediment should be considered. For assessing metal bioavailability and toxicity in sediment, the sequential extraction technique should be used in conjunction with other approaches. Metals linked with AVS may be liberated to sediments due to oxidation, storms, and dredging, among other things, and may impact aquatic environment [26]. As a result, bioavailable heavy metal concentrations rise, disrupting sedimentation dynamics. Different criteria of ΣSEM-AVS as benchmarks are proposed (Table 4.2).

4.5 C_{SEM}-Based Risk Quotient Model to Assess Metal Bioavailability in Sediments

Heavy metal bioavailability in sediments is linked to their toxicity, which is significant in evaluating sediment quality. The bioavailability of the metals through risk quotients were calculated by CSEM/(SQGVs) (Zn, Ni, Cd, Pb, and Cu). The use of CSEM measurement to estimate the risk induced by metals in aquatic systems was shown to be beneficial. The bioavailability of metals in oxic/suboxic sediments with an AVS shortage is highly influenced by sediment particle size, organic carbon

Table 4.2 Benchmarks of AVS-SEM models and their criteria [3, 14, 28, 32, 33, 37]

AVS-SEM models	Criterion for classification	Classification	References
ΣSEM/AVS	ΣSEM/AVS > 1	Acute toxicity may be generated	[3]
	ΣSEM/AVS < 1	Acute toxicity is less likely to occur	
ΣSEM/AVS	ΣSEM/AVS > 8.32	High macro-invertebrate toxicity	[37]
	2 > ΣSEM/AVS < 8.32	Occasionally toxic	
	ΣSEM/AVS < 2	No toxicity	
	ΣSEM-AVS/f_{oc} > 150	Toxicity may be occurred	
ΣSEM-AVS	ΣSEM-AVS > 0	Potential metal toxicity to sediment-dwelling organisms	[14]
	ΣSEM-AVS < 0	No toxic effects are expected to occur	
ΣSEM-AVS	ΣSEM-AVS > 5	probable associated adverse effects on aquatic life	[32]
	ΣSEM-AVS = 0–5	Possible associated adverse effects on aquatic life	
	ΣSEM-AVS < 0	No indication of associated adverse effects	
ΣSEM-AVS/foc	ΣSEM-AVS/f_{oc} > 3000	Adverse effects may be expected	[33]
	3000 > ΣSEM-AVS/f_{oc} > 130	Uncertain adverse effects	
	The AVS concentrations in the studied lakes vary clearly with the depth in the sediment cores/f_{oc} < 130	Not expected adverse effects	

(OC), and iron, manganese oxyhydroxides, [35, 47]. However, the bioavailability of various metals varies based on sediment characteristics. For example, the most critical variables influencing the bioavailability and toxicity of nickel in freshwater sediments are documented as AVS and iron [48, 49]. Organic carbon, on the other hand, had a critical effect on copper toxicity and bioavailability [50–52].

$$\mathbf{RQ_{SEM-TEL}} = \sum (\mathbf{C_{SEM}/TEL}) \tag{4.21}$$

$$\mathbf{RQ_{SEM-PEL}} = \sum (\mathbf{C_{SEM}/PEL}) \tag{4.22}$$

Threshold effect level (TEL) and probable effect level (PEL) are Chinese SQGVs for the protection of freshwater benthic organisms. To explore the connections between concentrations, quotients, and toxicities, regression analyses were used. TEL (mg/g) values for cadmium, copper, nickel, lead, and zinc are 3.02, 45.5, 31.4,

47.3, and 74.9, respectively, and PEL values are 18.9, 181, 76.9, 204, and 403 for the same metals [53, 54]. Zhang et al. [35], found that metal toxicity in benthic species such as *C. kiiensis*, M. *limosus*, and *C. tentans* changed based on feeding and burrowing. Toxicity was primarily due to solid-phase ingestion, which was linked with CSEM. Based on the findings of this investigation, integration between SQGVs (TEL and PEL) and CSEM proved to be a strong tool for identifying the risk associated with heavy metals in the sediments of Taihu Lake.

In previous studies, risk quotients (RQTEL; total metal/TEL and RQPEL; total metal/PEL) calculated by normalizing total heavy metal content in sediment with TEL and PEL (RQTEL; total metal/TEL and RQPEL; total metal/PEL) indicated a significant positive association with toxicity (*p* 0.001) [54].

Zhang et al. [35], performed toxicity test to three benthic creatures *M. limosus, C. kiiensis and C. tentans for* Autotomy and Mortality endpoint by evaluating relationships between toxicity and risk quotients defined upon $RQ_{SEM\text{-}TEL}$ and $RQ_{SEM\text{-}PEL}$. All organisms ate silt and burrowed into the top three centimeters of the sediment, which served as a source of hazardous metals. Different equations are set up achieved significant linear relationships [For *M. limosus* (Autotomy), Toxicity = 7.11 + 6.50 × $RQ_{SEM\text{-}TEL}$; Toxicity = 6.69 + 25.4 × $RQ_{SEM\text{-}PEL}$ 0.355, 0.093 and 0.371, 0.086 respectively] and [For C. *kiiensis* (Mortality), Toxicity = 5.51 + 13.3 × $RQ_{SEM\text{-}TEL}$; Toxicity = 5.01 + 51.4 × $RQ_{SEM\text{-}PEL}$ with r^2 and *p* 0.916, 0.001 and 0.927 0.001 respectively], [*For C. tentans* (Mortality) Toxicity = 9.98 + 6.61 × $RQ_{SEM\text{-}TEL}$; Toxicity = 9.44 + 26.0 × $RQ_{SEM\text{-}PEL}$ with r^2 and *p* 0.699, 0.006 and 0.729, 0.004]. While every two RQSEM-TEL, RQSEM-PEL, RQTEL, and RQPEL (RQ pairings) were highly linked (r^2 = 0.64–0.99, n = 27, *p* 0.01) [35].

At equilibrium, Copper is supposed to react with AVS preferentially, displacing all other metals. The remaining metals will react in the following order if the available AVS is not entirely saturated by copper: lead, cadmium, zinc, and nickel [55]. According to this model, the amount of copper in the sediment that is potentially bioavailable and harmful is:

$$Cu_b = (Cu_{SEM} - AVS) * (MW_{cu}) \tag{4.23}$$

where, Cu_b: stands for bioavailable copper concentration (mg/kg). Cu_{SEM}: molar concentration (moles/kg) as determined by simultaneous extraction. AVS: is AVS molar concentration (moles/kg). Copper molecular weight (mg/moles) denoted by MWcu.

4.6 Three Integrated Approaches: Eqpa, AVS and Sequential Extracted Technique Set by [56]

An equilibrium partitioning (EqPA) was built on the basis of three empirical hypotheses [57], inclusive:

1. Pollutants are transferred between the interstitial water phase and the sediment phase fast and reversibly, and the process eventually reaches equilibrium.
2. The amount of pollutants in the interstitial water is a major determinant of sediment toxicity.
3. Benthic organisms are sensitive to pollutants in the same way that aquatic organisms are.

The EqPA performed empirical approach with two hypotheses: the bioavailability of target metal in sediments and its biological effect for aquatic organisms [58] and heavy metal concentrations in interstitial water did not exceed WQC criterion for that metal [59, 60]. Using an equilibrium partitioning approach, numerical quality has emerged as a criterion for heavy metals [61, 62]. Kp revealed the link between concentration in the sediment (ρ_S) and interstitial water (ρ_{IW}).

$$\mathrm{KP} = \rho_S/\rho_{IW} \tag{4.24}$$

By setting ρ_{IW} equal to the water quality concentration; WQC for a given metal. The congruent sediment "safe" level, sediment quality criterion (SQC), could be evaluated using the following equation;

$$\mathrm{SQC} = \mathrm{K_P} \times \mathrm{WQC} = \rho_S \tag{4.25}$$

The BCR sequential extraction approach was used to classify four metal fractions in sediment with heavy metal contamination as extractable, reducible, oxidizable, and residual. The summation of lrachable fractions [extractable + reducible + oxidizable] in the sediment would flux with interstitial water, bioavailable metals, and harmful metals in the following manner correlated to prospective bioavailability in sediment (ρ_S). Portion of metals in the acid volatile sulfide ($[M]_{AVS}$) did not undergo equilibrium process, and residual metal fractions ($[M]_R$) from metal fractionation technique did not undergo partitioning equilibrium procedure with interstitial water either since bivalent metal (Me^{2+}) and bivalent sulfide (S^{2-}) react together to form sulfide precipitates [9].

As a result, the SQC calculation formula was rewritten as follows:

$$\mathrm{SQC} = \mathrm{K_P WQC} + [M]_R + [M]_{AVS} \tag{4.26}$$

where, KP is the partition coefficient; WQC [33] is water quality criterion (μg/l); $[M]_R$ is the residual fraction of metal linked to the crystal lattice in sediments (μg/g); $[M]_{AVS}$ is portion of metal bound to sulfide in sediments (μg/g).

4.7 Model Describing Rapid Sulfide-Buffering Capacity

Azzoni et al. [63], described the rapid sulfide-buffering capacities (βL) which is represented by a potential buffering capability for the trapping of S^{2-} by reaction with Fe (III).

$$\beta L = [LFe\,(II) - AVS] + 1.5LFe\,(III) \tag{4.27}$$

$$3\,S^{2-} + 2Fe(III) = S^{0} + 2FeS \tag{4.28}$$

[LFe(II) – AVS] is the buffering capacity of Fe^{2+} (not sulfurized). LFe^{3+} is the buffering capacity of labile Fe^{3+} oxides that immediately absorb dissolved sulfide in accordance with the 3:2 stoichiometry between S^{2-} and FeS (3:2) in Eq. (4.28). High buffering saturation likewise mitigates the negative effects on benthic ecosystems [36, 64, 65]. AVS/LFe ratios less than 0.25 signify a high level of residual buffering, but values between 0.75 and 1 signify exhausted buffering. According to [36, 66], it is anticipated that the activity of sulfide created by sulfate reducing bacteria will be released when AVS/LFe is larger than 1.

4.8 Structural Equation Modelling (Cutting-Edge Modeling Technique)

Structural equation modelling was used to examine the simultaneously extracted metals/acid volatile sulfides (SEM/AVS) index's capacity to assess the environmental risk from potentially harmful constituents. The top-performing model highlighted the key soil (sediment) characteristics that influence AVS accumulation and SEM speciation in these soils. AVS and SEM effect charts produced by piecewise structural equation modelling using partial linear mixed effects models. The structural equation modeling technique was performed according to [67], using [R software Ver 3.5.1; piecewise SEM package Ver 2.0.2 (R Development Core Team 2018). The model was run using based on the hypotheses that Flooding directly influences the buildup of AVS, and Fe oxides, organic carbon, and carbonates may also function as SEM's binding aspects.

The results of structural equation modelling revealed a highly substantial effect of carbonates and labile organic carbon on SEM and a strong and favorable influence of pH and carbonates on AVS. To identify the potential contributions of labile organic carbon or carbonates as alternative binding phases, individual SEM components were also taken into consideration separately. Cu, Ni, and Zn that were simultaneously removed were more closely bonded to carbonates [68] (Fig. 4.9).

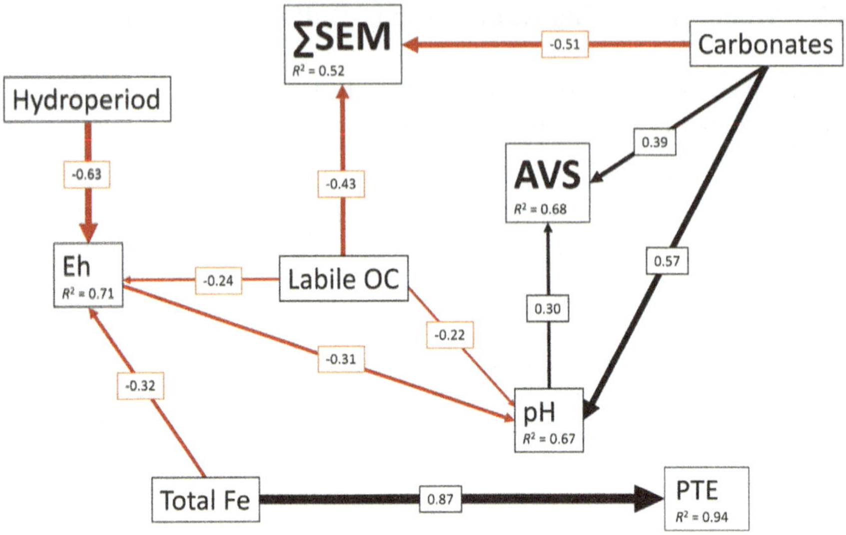

Fig. 4.9 The tested structural equation modeling's outcomes. The arrows in black and red denote statistically significant positive and negative associations, respectively. The size of the effect, as reported in the text box, is inversely correlated with the arrow width. Non-reportable relationships are not considered. Additionally displayed within boxes are conditional R^2 values. PTE is for a potentially hazardous element [68]

4.9 Statistical Approaches

Since sulfide models are viewed as excellent ecological risk tools in sediment assessments with a significant reliance on statistical analysis, several researchers applied statistical analyses to AVS-SEM models, which are of major value in directing the toxicity assessment of sediments [19, 69–71].

4.9.1 Correlation Analysis

Fang et al. [26], illustrated the results of the Pearson correlation coefficient in three sediment core samples from Pearl River Estuary, South China (Table 4.3) which confirmed the strong relationship between metals and sulfide parameters. The SEM levels accounted for a relatively large percentage of the total metal concentrations.

Gao et al. [72], offered another viewpoint for metal sulfide models. As shown in Table, their findings showed that the analyzed trace metals had substantial correlations with one another, indicating a common source. According to their significant correlations with the environmental parameters, the grain size structure, TOC, and water content were all major contributing factors for each of the SEM components.

Table 4.3 Person correlation coefficients of total metal (Cu, Ni, Pb, Zn and Cd) to SEM and AVS to Fe and Al in the sediment cores ($P < 0.05$) [26] with permission

Sample	N	Critical r	R						
			AVS-Fe	AVS-Al	Cd	Cu	Ni	Pb	Zn
Core1	34	0.341	0.5799	– 0.3284	0.93585	0.419	0.343	0.9646	0.95
Core2	12	0.566676	-0.3004	– 0.191	– 0.0786	0.6978	0.787	0.235	0.846
Core3	21	0.433	0.55597	0.59185	– 0.467	-0.017	NA	0.338	0.42655

This was also true for the pH condition for all five metals, with the exception of Cd,whereas Eh appeared to have no effect on any of them (Table 4.4).

4.9.2 *Multiple Regression Analysis as a Powerful Tool to Predict Sediment Quality of Marine Environment Based on AVS, Metals, and Organic Matter*

Zhang et al. [5], used multiple regression analysis to create equations that described ecological risks and predicted the toxicity of heavy metals (Zn, Cu, Cd, Pb, and Ni) in the sediments of China's Haihe River. Sediment properties used in the analysis included AVS, TOC, and PSD; particle size distribution. Using multiple regressions, the risk quotients estimated from total concentrations and SQG values were modified. The sediment toxicity was examined using tubificids and chironomids as benthic species, and the connections between observed toxicity and adjusted risk quotients were evaluated. The sum of the risk quotients for each metal for C_{Sed} (Q) was used to calculate the risk quotients by Eq. (4.29).

$$Q = \sum C_{sediment}/SQG \tag{4.29}$$

$$\text{Microtox}^{\circledR}\ \text{toxicity} = k \cdot \lg Q * + b \tag{4.30}$$

$$\text{Modified TQ}(Q*) = Q \cdot (AVS/AVS_{Ref})^{\textbf{slope Avs}} \times (TOC/TOC_{Ref})^{\textbf{slope TOC}} \times (RSD/RSD_{Ref})^{\textbf{slope RSD}} \tag{4.31}$$

$$\text{Toxicity} = k \cdot \big(\lg Q + \text{Slope}_{TOC} \cdot \lg TOC + \text{Slope}_{AVS} \cdot \lg AVS + \text{Slope}_{PSD} \cdot \lg PSD\big) + b^{*} \tag{4.32}$$

The level of heavy metal in the sediment is has the symbol (C). Sediment quality guideline value is expressed as SQG. Metals, bioavailability in sediments is regulated by the physico-chemical characteristics of the sediments, hence Q* is derived by

Table 4.4 Spearman correlation coefficients of AVS/SEM with geochemical sediment features based on [72] with permission

	[AVS)	[SEM]	Cd	Cu	Ni	Pb	Zn	pH	Eh	%Clay	%Silt	% Sand	%TOC	%Water content
[AVS]	1	0.259	– 0.221	0.171	0.310	0,235	0.271	0.172	– 0.468[c]	0.278	0180	– 0.210	0.130	0.018
[SEM]	0.259	1	0.564[b]	0.968[a]	0.961[a]	0.988[a]	0.991[a]	– 0.483[c]	– 0.302	0.869[a]	0.873[a]	– 0.902[a]	0.856[a]	0.790[a]
Cd	– 0.221	0.564[b]	1	0.573[b]	0.478[c]	0.564[b]	0.567[b]	– 0.375	0.143	0.401c	0429[c]	– 0037′	0.597[b]	0.601[b]
Cu	0.171	0.968[a]	0.573[b]	1	0.445[c]	0.95[a]	0930[a]	– 0.507[b]	– 0.323	0.888[a]	0.856[a]	– 0.893[a]	0.869[a]	0.7738[a]
Ni	0.310	0.961[a]	0.478[c]	0.445[c]	1	0.922[a]	0.984[a]	– 0.531[b]	– 0.380	0.885[a]	0.890[a]	0.918[a]	0.798″	0.7298[a]
Pb	0,235	0.988[a]	0.564[b]	0.95[a]	0.922[a]	1	0.985[a]	– 0.443[c]	– 0.281	0.847[a]	0834[a]	– 0.866[a]	0.857[a]	0.7888[a]
Zn	0.271	0.991[a]	0.567[b]	0930[a]	0.984[a]	0.985[a]	1	– 0.445[c]	– 0.262	0.829[a]	0.848[a]	0. 872[a]	0.8408[a]	0.7878[a]

[a] $P < 0.001$
[b] $0.001 < P < 0.01$
[c] $0.01 < P < 0.05$

changing Q with sediment parameters as TOC, AVS, and PSD. The C sediment gives the total amount of each heavy metal present in the sediments. The term SQG stands for Sediment Quality Guideline Value. The bioavailability of metals in sediments is influenced by the physico-chemical features of the sediments, hence Q* is calculated by adjusting Q with sediment characteristics like AVS, TOC, and PSD. It is believed that lgQ* and benthic species' toxicity are positively associated. Zhang et al. [5], arrived at Eq. (4.32) by rearranging Eqs. (4.29), (4.30), and (4.31). As a result, toxicity is the proportion of toxic outcomes, such as the emergence or demise of the organisms. Specifically, the slopes are for TOC, AVS, and PSD. The reference values for the characteristics of appropriate reference sediment are AVS^{Ref}, TOC^{Ref}, and PSD^{Ref}. PSD of 50% (63 m), AVS of 0.005 mol/g, and TOC of 1% were the reference values. Through the use of linear regressions, the constants k, b, and b* were determined. The toxicity testing methods were disclosed by [35, 54]. Ju et al. [73], evidenced that Microtox® toxicity was significantly positively correlated with different metal pollution indices; PLI (pollution load index), mCD (modified degree of contamination).

4.9.3 Principal Component-Factor Analysis

Wang et al. [21], applied PCA-factor analysis on the entire datasets of 22 variables in sediment samples from Maluan Bay, China included: sediment variables (total metal, SEM and AVS), environmental parameters (pH, Eh, water content, TOC and grain size distribution), and normalized metal concentrations (SEM/AVS, SEM-AVS, (SEM-AVS)/fOC). The table lists three extracted factors accounted for about cumulative co variance 85% calculated through (Varimax-rotated) with factor loadings to each variable Table 4.5. The high communalities showed that the newly chosen factors adequately describe all of the studied variables. F1 showed 40.5% of the total variance with high loading values for TOC, [SEM], total Zn, SEM_{Zn} and SEM_{Cu} and medium to SEM-AVS)/*fOC*, SEM/AVS and SEM-AVS. F2 accounted for 30.2% of variance with positive loadings for pH, total Ni, SEM_{Ni}, total Pb, SEM_{Pb} and total Cu and negative loadings for SEM_{Cd} and silt. Strong relationships with this factor were demonstrated by Ni and Pb. F3 accounted for 14.3% of the initial variation and had controlled by high negative factor loadings loading to TOC, silt, and total Cd in, with the highest positive factor loadings of Eh and moisture content (MC) and the medium positive factor loadings of SEM/AVS, SEM-AVS, and (SEM-AVS)/*fOC*. In conclusion, [21], supported the integrated use of a PCA-F to characterize the pollutants for sustainable coastal management.

Table 4.5 Varimax rotated PCA-FA for 22 variables included AVS/SEM model parameters in Maluan Bay, China (sediment samples) [21] with permission

	Factor loading		
	Factor 1	Factor 2	Factor 1
% Variance explained	40.5	30.21	14.27
% Cumulative variance	40.5	0.71	84.98
Initial eigen value	8.91	6.65	3.14
pH	– 0.29	**0.81**	0.15
Eh	– 0.16	0.17	**0.88**
MC	0.07	0.11	**0.98**
TOC	**0.65**	– 0.05	– *0.47*
%Clay	– 0.26	0.21	0.21
%Silt	0.24	– *0.33*	– *0.53*
%Sand	– 0.11	0.40	**0.84**
Cd-$_{TM}$	– 0.36	0.43	– *0.61*
SEM_{Cd}	– 0.12	– *0.52*	0.10
Cu-$_{TM}$	**0.53**	**0.77**	– 0.10
SEM_{Cu}	**0.77**	**0.6**	– 0.01
Ni-$_{TM}$	0.04	**0.91**	0.25
SEM_{Ni}	0.5	**0.72**	0.34
Pb-$_{TM}$	0.18	**0.96**	0.11
SEM_{Pb}	**0.54**	**0.8**	0.13
Zn-$_{TM}$	**0.93**	0.11	– 0.07
SEM_{Zn}	**0.9**	– 0.11	– 0.12
SEM	**0.94**	0.22	– 0.04
AVS	0.36	0.06	– *0.76*
[SEM]/[AVS]	**0.61**	0.42	**0.57**
[SEM]-[AVS]	**0.67**	0.18	**0.67**
[SEM]-[AVS]/*foc*	**0.6**	0.36	**0.62**

Bold number; positively loaded
Italic number; negatively loaded
underline Positive loading indicated that the parameter was directly proportional to the factor extracted by principal component analysis, rotation method, Varimax with Kaiser normalisation. However, negative loading revealed that the parameter inversly proportion to the factor.

4.10 Approval of Various Assessment Techniques and Their Environmental Implications

He et al. [28], used various techniques to assess the potential toxicity of total metals to aquatic biota: ERL, Effect Range Low and ERM; Effect Range Median (USA), TEL; threshold effect level, PEL; probable effect level (Australia and New Zealand) and

AVS-SEM models, respectively. These SQGs specify concentration ranges where unfavorable biological effects are considered infrequently (below the low-range value), rarely (between the low and mid-range values), and frequently (between the low and mid-range values) (above the mid-range value). Due to the limitations of each approach, the results were inconclusive. The SQGs that were designed for a given location were affected by geography and environmental conditions [59, 74, 75].

The effects of particle size, pH, particular species, exposure period, and population stress on SQG susceptibility, bioavailability, and toxicity were not extensively investigated [76, 77]. The significance of AVS models in estimating harmful effects on aquatic species has been highlighted (Van Griethuysena et al. 2003). Metals from sediments may not cause toxicity to biota even if $\sum$SEM levels are significantly greater than AVS values [15, 45]. Some geochemical processes may be able to remove metals from pore water, reducing their bioavailability. Complexation, precipitation, metal ion adsorption, and dissolution with carbonates, clays, and/or oxyhydroxide minerals are some of the processes involved. AVS was also influenced by a range of dynamic biochemical processes, including oxidation, deposition, and bioturbation, causing temporal and geographical variations in AVS [76, 78, 79]. To specify a single technique to forecast sediment toxicity may not be sufficient. Because it is a simple operation with high efficiency for ecotoxicological evaluation, the addition of AVS-SEM criteria to assess sediment quality will be required [30].

The AVS-SEM models, for example, can assist in identifying regions of considerable environmental concern, allowing priority locations to be created for further comprehensive research [80]. Because a single method of toxicity evaluation may be insufficient, the integration of sediment quality guidelines, SQGs, AVS-SEM models, and fractionation techniques will be more compelling in terms of the actual decision making of sediment quality management strategies in the aquatic environment. In the next investigation, biological effect evaluations will be necessary for addition to chemical analyses.

4.11 Chemical Kinetic Models for AVS and SEM Dissolution

The obvious definition is that the reaction rate is proportionate to the surface area and has a rate constant, K. A more specific definition is explained in the equation:

$$dV(t) = -K\, A(t)dt \tag{4.33}$$

dV(t) is the amount of dissolved sulfide mineral at time t after dt. It is assumed that at any time A, and it is proportional to the surface area (m^2) at time t. Kinetic dissolution of AVS and SEM are interpreted chemically by fitting to a shrinking particle model (SPM). The SPM is a mechanistic model in which the rate of dissolution is directly proportional to the surface area. It expresses a knowledge of the physical and chemical procedures necessary to dissolve a solid particle [81]. Many researchers have noted that an SPM and a first-order model (FOM) may simulate AVS extraction equally well [81, 82].

K has the units of volume (m^3) per unit time, per unit volume reacted and is affected by pH, temperature, oxygen concentration, and ionic strength.

Following SPM model, it will be demonstrated that $K = K_{spm}$

$$C_t = C_{max} - C_{max}\left(1 - K_{spm}t\right)^3 \tag{4.34}$$

where Ct is the concentration at time t, C_{max} is the maximum AVS or SEM metal concentration that can be extracted from the sediment, and k SPM is the reaction rate constant.

Apply It expresses a knowledge of the physical and chemical procedures necessary to dissolve a solid particle [81]. Many researchers have noted that an SPM and a first-order model may simulate AVS extraction equally well.

As a semiempirical alternative to the SPM, FOM was used as a First-order reaction to describe kinetics which can approximate many natural processes.

$$C_t = C_{max}(1\_e^{k_{FOM}t}) \tag{4.35}$$

Pyrite dissolution and AVS oxidation have both been explained using the equation above. This method might be applied as a prescreening technique for risk or sediment site characterization. It is doubtful that SEM metals are present as discrete MeS when AVS $\approx$ SEM since, AVS and SEM metals dissolving patterns differ in their bath ways. As a result, a correction for AVS in such a site's risk assessment may be questionable [81, 83].

4.11.1 Deep Insight into Kinetic Reactions of Cd-AVS Through Solubility and Displacement

In sediments, sulfides exist mainly as iron mono sulfides such as mackinawite, greigite, iron bisulfide, and pyrite. AVS is related to iron and manganese monosulfides, which are more easily soluble. In the cold acid extraction employed to quantify AVS, the more stable iron pyrite and organic sulfide associated with organic materials in sediments are insoluble [84]. The role of sulfide as a regulating ion in metal concentrations in marine sediment and interstitial water is well established [3–5, 27, 45, 69, 85–87]. The aqueous phase sulfide and iron mono sulfide, FeS (s), are in equilibrium: The following equations explain what happens when cadmium is introduced to the aqueous phase:

$$FeS(s) \leftrightarrow Fe^{2+} + S^{2-} \tag{4.36}$$

$$Cd^{2+} + S^{2-} \leftrightarrow CdS_{(s)} \tag{4.37}$$

$$Cd^{2+} + FeS(s) \leftrightarrow Cd^{2+} + Fe^{2+} + S^{2-} \tag{4.38}$$

$$Cd^{2+} + FeS(s) \leftrightarrow CdS(s) + Fe^{2+} \tag{4.39}$$

The displacement of Fe in iron mono sulfide FeS(s) with Me^{2+} to create less soluble metal sulfides is the chemical mechanism of AVS binding to metal ions [24, 25, 87, 88]. Cadmium's LC_{50} in marine sediments is determined by the sediment phase AVS. For the same total amount of metal, sediments with various characteristics will have varying degrees of toxicity. The organism's response can be understood by developing a link between the levels of metal in the interstitial water and the sediments. It's crucial to consider the sediment properties that govern how much metal is distributed across the solid and aqueous phases. According to certain theories, iron and manganese oxides, organic carbon, and other substances may have a significant role in controlling the toxicity of metals in sediments [56].

4.12 AVS/Fractionation Techniques

Many researchers investigated SEM in line with metals extracted by sequential extraction techniques as ecological risk tools, which are of great significance in guiding the toxicity assessment of sediments [19, 69–71]. According to [71], instead of the traditional simultaneous extraction of heavy metals (SEM)/AVS, which would overestimate the toxicity of heavy metals, the ratio of (acid-soluble fraction (F1), reducible fraction (F2), + oxidizable fraction (F3)/AVS was the appropriate index for ecological risks associated with heavy metals. Li et al. [71] investigated heavy metals (Pb, Cu, Cd, and Zn) and sulfur in environmental sediments (the North Yellow Sea). The Pearson correlation analysis was conducted for AVS, RIS; reactive inorganic sulfide, CRS, chromium (II)-reducible sulfur (CRS), and metal speciation; BCR (F1 + F2 + F3 + F4); and elemental sulfur; ES (Fig. 4.10).

This study revealed that AVS/RIS was positively correlated fractionation phases of metals in sediment. These correlations emphasize that AVS to CRS primarily occur through the H_2S pathway ($FeS + H_2S = FeS_2 + H_2$) and polysulfide pathway ($FeS + S^0 = FeS_2$) Both approaches require ES, which is frequently formed by incomplete oxidation of RIS by Mn/Fe (oxide) [34].

According to [89], the findings of the AVS and SEM have contributed to the elucidation of the anomaly seen in the alteration of heavy metal speciation patterns following oxidation of the anoxic sediment sample. In light of this discovery, AVS and SEM results should be added to the sequential extraction process used to determine the heavy metal speciation patterns of anoxic soil samples.

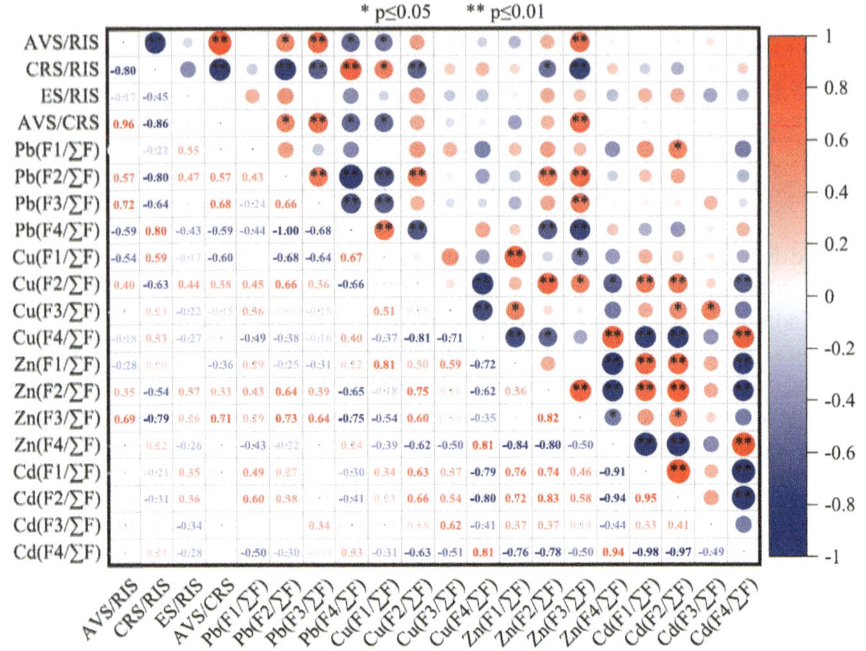

Fig. 4.10 Correlation analysis for the RIS, AVS, ES, CRS and the geochemical fractions of Zn, Pb, Cu, and Cd in sediments (North Yellow Sea) [71] with permission

4.13 SEM/Total Metals

According to [26], the SEM/total metal ratios indicated that there were various interactions with the sulfide phase and/or various solubilities during the HCl-extraction. The heavy metal reactivity in the sediments was also evident through this ratio. Chai et al. [90] designed a study to highlight the reactivity of the heavy metals in the sediments. A deep insight into SEM and the total metals for the 16 sites revealed that all SEM were significantly correlated with total metals, with different degrees referring to the mode of action of each metal (Fig. 4.11). As shown in Table 4.6, the SEM concentrations and their contributions to $\sum$SEM (R1) and total metal concentration (R2). The results revealed that the four-acid extracted-metal concentrations are differed significantly. 69.7–94.2% for SEM-Zn of $\sum$SEM, however, SEM-Cd (more toxic) had 1%. SEM metal, hold to be a large percentage of the total metal concentrations in most sites, SEM-Pb (59.8–96.5%), SEM-Cd (47.2–81.0%), SEM-Cu (5.2–65.6%), and SEM-Zn (24.4–57.1%). the lower values of R2 for Cd, Cu and Zn, [91], revealed that to different binding phases (Fe/MnOx and organic matter). Additionally, the SEM-metal/total metal ratio varied by location and showed no consistent patterns, which may be attributed to the different locations with different degrees of sediment heterogeneity.

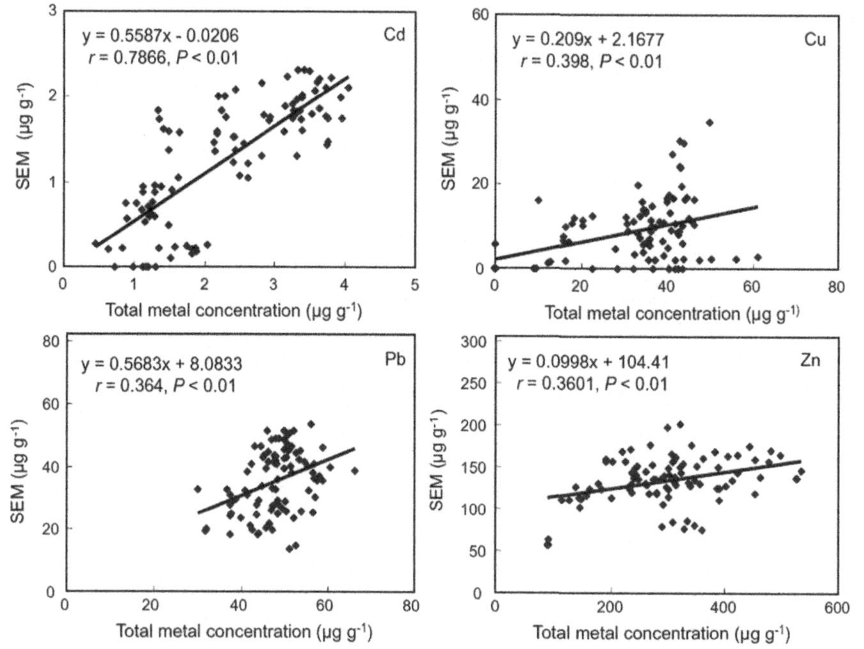

Fig. 4.11 Total metals (Cd, Cu, Pb and Zn) and corresponding SEM in surface sediments of Futian mangrove forest, China [90] with permission

4.14 Bioassay

Li et al. [92], designed a field investigation of the correlation between SEM-AVS models and the bioaccumulation of metals by *Limnodrilus sp.* in a highly polluted river (Foshan Waterway, Guangdong, China.). The majority of the research region was under anaerobic and severely reducing conditions, and trace metal concentrations were high in surface sediments. AVS and concurrently extracted metals (SEM; average AVS $= 20.3$ μmol g^{1-}, average $SEM_{(Cd+Zn+Ni+Pb+Cu)} = 9.42$ μmol g^{1-}) concentrations were also high. The predominant species was *Limnodrilus sp.* There was no significant relation between BIO5 and SEM-AVS, and the absolute value of BIO5 was high both in stations where SEM-AVS was positive and in stations where it was negative (average BIO5 $= 4.07$ μmol g^{1-}), indicating that there was no direct link between SEM-AVS and metal accumulation in *Limnodrilus sp.* (Fig. 4.12). This was probably due to the fact that *Limnodrilus sp.* eat sediment grains as their primary food supply, which means that pore water metals have no effect on metal bioaccumulation. However, BIO5 was significantly correlated with SEM5 ($r = 0.795$, $p = 0.01$) (Fig. 4.13), indicating that the high concentration of SEM in sediments may be a significant factor in metal accumulation in *Limnodrilus* species, which can assimilate SEM with the aid of digestive fluids. This phenomenon might be explained by the dependence of benthic creatures' metal buildup on ecology and eating habits.

Table 4.6 SEM, %$\sum$SEM and total metal concentrations % [90] with permission

Site	Cd			Cu			Pb			Zn		
	Concentration	R_1	R_2	Concentration	R_1	R_2	Concentration	R_1	R_2	Concentration	R_1	R_2
1	2.11 ± 0.16	0.7	62.6	21.60 ± 5.34	11.8	52.0	41.64 ± 3.66	7.2	87.2	147.79 ± 11.21	80.4	31.0
2	1.91 ± 0.22	0.7	57.6	14.08 ± 4.34	9.0	36.9	28.59 ± 3.08	5.7	68.1	133.05 ± 11.86	84.6	47.5
3	1.72 ± 0.24	0.7	52.0	12.17 ± 1.99	4.0	38.9	31.58 ± 4.24	6.5	73.3	134.47 ± 9.36	88.8	51.6
4	2.17 ± 0.13	0.8	58.4	13.83 ± 2.27	8.7	39.4	23.91 ± 5.72	4.6	69.6	138.30 ± 10.73	85.9	53.7
5	1.82 ± 0.07	0.7	53.6	9.07 ± 3.00	6.4	34.7	28.58 ± 3.18	6.3	59.8	122.93 ± 6.76	86.5	55.1
6	0.92 ± 0.10	0.3	65.6	6.03 ± 2.45	3.9	17.5	45.10 ± 4.28	9.1	96.5	134.75 ± 11.31	86.6	50.4
7	1.62 ± 0.16	0.6	61.3	5.38 ± 1.56	2.0	13.6	41.79 ± 5.27	8.9	84.9	130.26 ± 18.01	88.4	41.3
8	1.66 ± 0.20	0.5	47.6	8.16 ± 1.99	4.6	20.8	43.87 ± 3.62	7.6	89.9	157.27 ± 13.57	87.2	43.1
9	1.49 ± 0.08	0.5	65.9	12.48 ± 2.18	6.6	29.0	48.79 ± 2.74	8.0	96.0	163.43 ± 4.89	85.0	39.0
10	1.87 ± 0.21	0.5	81.0	12.78 ± 2.91	6.5	32.1	42.58 ± 3.03	6.7	78.4	174.78 ± 23.63	86.3	57.1
11	1.21 ± 0.13	0.6	47.2	20.09 ± 12.04	16.7	65.6	47.64 ± 3.95	13.0	94.9	79.99 ± 4.65	69.7	24.4
12	0.60 ± 0.09	0.3	50.4	2.51 ± 0.30	0.9	5.2	35.19 ± 3.61	8.0	61.6	124.97 ± 9.39	90.8	30.6
13	0.75 ± 0.15	0.3	68.0	3.75 ± 2.52	2.4	9.5	40.36 ± 2.56	8.2	78.5	138.84 ± 12.45	89.2	38.9
14	0.19 ± 0.06	0.1	16.1	0.00 ± 0.00	0.0	-	20.11 ± 1.08	7.5	45.0	87.18 ± 31.87	92.3	79.9
15	0.23 ± 0.02	0.1	13.3	1.58 ± 0.20	0.7	13.6	19.60 ± 4.64	5.0	42.3	114.73 ± 8.19	94.2	74.6
16	0.00 ± 0.00	0.0	0.0	8.00 ± 2.01	2.3	90.7	26.28 ± 1.28	4.8	52.9	160.72 ± 7.32	92.9	75.0

Metal concentration showed as: mean concentration ± SD (mg/kg); R1 = %SEM-metal/$\sum$SEM; R2 = % SEM-metal/total-metal

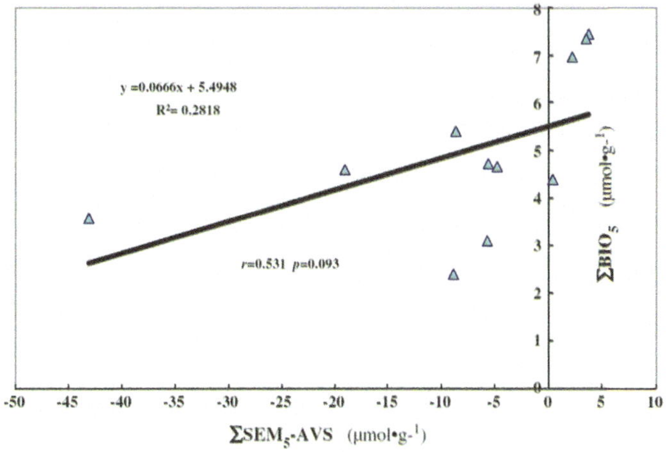

Fig. 4.12 (∑BIO_5) against (∑SEM5-AVS) in *Limnodrilus sp* [92]

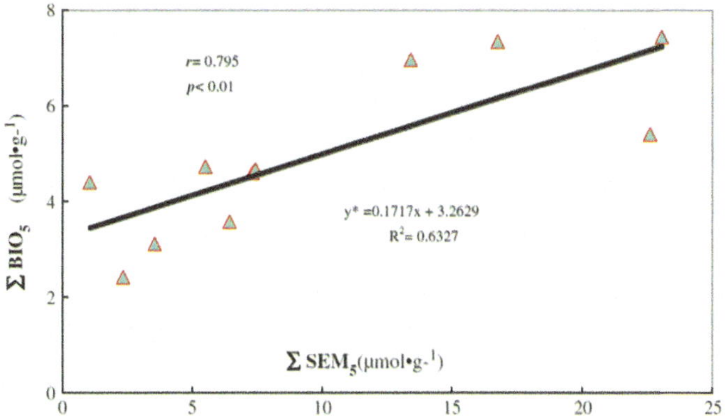

Fig. 4.13 (∑BIO_5) relevant to (∑SEM_5) in *Limnodrilus sp.* [92]

Limnodrilus species use sediment particles as their primary food source and reside in the mud at the bottom of aquatic systems. Fluids in *Limnodrilus* digestive tracts could potentially have a role in the substantial correlation between metal accumulation and the ability to absorb SEM.

References

1. Simpson SL, Rosner J, Ellis J (2000) Competitive displacement reactions of cadmium, copper, and zinc added to a polluted, sulfidic estuarine sediment. Environ Toxicol Chem 19(8):1992–1999. https://doi.org/10.1002/etc.5620190806
2. Simpson SL (2001) A rapid screening method for acid-volatile sulfide in sediments. Environ

Toxicol Chem 20(12):2657–2661. https://doi.org/10.1002/etc.5620201201
3. Di Toro DM, Mcgrath JA, Hansen DJ, Berry WJ, Paquin PR, Mathew R, Wu KB, Santore RC (2005). Predicting sediment metal toxicity using a sediment biotic ligand model: methodology and initial application. Environ Toxicol Chem 24(10):2410. https://doi.org/10.1897/04-413r.1
4. Meysman JRP, Middelburg JJ (2005) cid-volatile sulfide (AVS)—a comment. Mar Chem 97:206–212. https://doi.org/10.1016/j.marchem.2005.08.005
5. Zhang Y, Spadaro DA, King JJ, Stuart SL (2020) Improved prediction of sediment toxicity using a combination of sediment and overlying water contaminant exposures. Environ Pollut 266:115187. https://doi.org/10.1016/j.envpol.2020.115187
6. Ribeiro AP, Figueiredo AMG, Santos JO, Dantas E, Cotrim MEB, Cesar Lopes Figueira R, Silva Filho EV, Cesar Wasserman J (2013) Combined SEM/AVS and attenuation of concentration models for the assessment of bioavailability and mobility of metals in sediments of Sepetiba Bay (SE Brazil). Mar Pollut Bull 68(1–2):55–63. https://doi.org/10.1016/j.marpolbul.2012.12.023
7. Ankley GT, Mattson VR, Leonard EN, West CW, Bennett JL (1993) Predicting the acute toxicity of copper in freshwater sediments: evaluation of the role of acid-volatile sulfide. Environ Toxicol Chem 12(2):315–320. https://doi.org/10.1002/etc.5620120214
8. Morse JW, Millero FJ, Cornwell JC, Rickard D (1987) The chemistry of the hydrogen sulfide and iron sulfide systems in natural waters. Earth Sci Rev 24:1–42. https://doi.org/10.1016/0012-8252(87)90046-8
9. Chapman PM, Wang F, Janssen C, Persoone G, Allen HE (1998) Ecotoxicology of metals in aquatic sediments: binding and release, bioavailability, risk assessment, and remediation. Can J Fish Aquat Sci 55(10):2221–2243. https://doi.org/10.1139/f98-145
10. Huerta-Diaz MA, Tessier A, Carignan R (1998) Geochemistry of trace metals associated with reduced sulfur in freshwater sediments. J Appl Geochem 13:213–233. https://doi.org/10.1016/S0883-2927(97)00060-7
11. Morse JW, Arakaki T (1993) Adsorption and coprecipitation of divalent metals with mackinawite (FeS). Geochim Cosmochim Acta 57(15):3635–3640. https://doi.org/10.1016/0016-7037(93)90145-m
12. Krouse HR, McCready RGL (1979) Reductive reactions in the sulfur cycle. In: Trudinger PA, Swaine DJ (eds) Bio-geochemical cycling of mineral-forming elements. Elsevier, New York, NY, pp 315–368. https://doi.org/10.1016/S0166-1116(08)71063-X
13. Morse JW, Cornwell JC (1987) Analysis and distribution of iron sulfide minerals in recent anoxic marine sediments. Mar Chem 22:55–69. https://doi.org/10.1016/0304-4203(87)90048-X
14. Di Toro DM, Mahony JD, Hansen DJ, Scott KJ, Hicks MB, Mayr SM, Redmond MS (1990) Toxicity of cadmium in sediments: the role of acid volatile sulfide. Environ Toxicol Chem 9(12):1487–1502. https://doi.org/10.1002/etc.5620091208
15. Di Toro DM, Mahony JD, Hansen DJ, Scott KJ, Carlson A, Ankley GT (1992) Acid volatile sulphide predicts the acute toxicity of cadmium and nickel in sediments. Environ Sci Technol 26:96–101. https://doi.org/10.1021/es00025a009
16. Islamoglu S, Yilmaz L, Ozbelge HO (2006) Development of a precipitation-based separation scheme for selective removal and recovery of heavy metals from cadmium rich electroplating industry effluents. Sep Sci Technol 41(15):3367–3385. https://doi.org/10.1080/01496390600851665
17. Kim BR, Gaines WA, Szafranski MJ, Bernath EF, Miles AM (2002) Removal of heavy metals from automotive wastewater by sulfide precipitation. J Environ Eng 1287:612–623. https://doi.org/10.1061/(ASCE)0733-9372(2002)128:7(612)
18. Helser J, Vassilieva E, Cappuyns V (2022) Environmental and human health risk assessment of sulfidic mine waste: bioaccessibility, leaching and mineralogy. J Hazardous Mater 424(Part A):127313. https://doi.org/10.1016/j.jhazmat.2021.127313
19. Diao F, Liu Y, Xu D, Zeng Q, Wang Z, Wang Y (2022) Comparison of acid volatile sulphide, metal speciation, and diffusive gradients in thin-film measurement for metal toxicity assessment of sediments in Lake Chaohu, China. Sci Total Environ 837:155438. https://doi.org/10.1016/j.scitotenv.2022.155438

20. Painuly AS, Shrestha S, Hackney P (2015) Bioavailability of heavy metals using simultaneously extracted metal/acid volatile sulfide in the sediments of lake Burragorang, NSW, Australia. J Environ Pollut Hum Health 3(1):12–17. https://doi.org/10.12691/jephh-3-1-3
21. Wang Z, Yin L, Qin X, Wang S (2019) Integrated assessment of sediment quality in a coastal lagoon (Maluan Bay, China) based on AVS-SEM and multivariate statistical analysis. Mar Pollut Bull 146:476–487. https://doi.org/10.1016/j.marpolbul.2019.07.005
22. He J, Lü C, Fan Q, Xue H, Bao J (2011) Distribution of AVS-SEM, transformation mechanism and risk assessment of heavy metals in the Nanhai Lake in China. Environ Earth Sci 64(8):2025–2037. https://doi.org/10.1007/s12665-011-1022-z
23. Schippers A (2004) Biogeochemistry of metal sulfide oxidation in mining environments, sediments, and soils. In: Amend JP, Edwards KJ, Lyons TW (eds) Sulfur biogeochemistry—past and present. Geological Society of America Special Paper 379, Boulder, Colorado, 49–62. https://doi.org/10.1130/0-8137-2379-5.49
24. Machado W, Villar LS, Monteiro FF, Viana LCA, Santelli RE (2010) Relation of acid-volatile sulfides (AVS) with metals in sediments from eutrophicated estuaries: is it limited by metal-to-AVS ratios? J Soils Sediments 10:1606–1610. https://doi.org/10.1007/s11368-010-0297-0
25. Nizoli EC, Luiz-Silva W (2012) Seasonal AVS–SEM relationship in sediments and potential bioavailability of metals in industrialized estuary, southeastern Brazil. Environ Geochem Health 34:263–272. https://doi.org/10.1007/s10653-011-9430-2
26. Fang T, Li X, Zhang G (2005) Acid volatile sulfide and simultaneously extracted metals in the sediment cores of the Pearl River Estuary, South China. Ecotoxicol Environ Safety 61(3):420–431. https://doi.org/10.1016/j.ecoenv.2004.10.004
27. McGrath JA, Paquin PR, DiToro DM (2002) Use of the SEM and AVS approach in predicting metal toxicity in sediments. Fact sheet on environmental risk assessment. Published by the International Council on Mining and Metals (ICMM). No. 10. London, UK
28. He Y, Meng W, Xu J et al (2015) Spatial distribution and toxicity assessment of heavy metals in sediments of Liaohe River, northeast China. Environ Sci Pollut Res 22:14960–14970. https://doi.org/10.1007/s11356-015-4632-2
29. Simpson SL, Batley GE (2007) Predicting metal toxicity in sediments: a critique of current approaches. Integr Environ Assess Manag 3:18–31. https://doi.org/10.1002/ieam.5630030103
30. De Lange HJ, van Griethuysen C, Koelmans AA (2008) Sampling method, storage and pretreatment of sediment affect AVS concentrations with consequences for bioassay responses. Environ Pollut 151:243–251. https://doi.org/10.1016/j.envpol.2007.01.052
31. Liu D, Wang J, Yu H et al (2021) Evaluating ecological risks and tracking potential factors influencing heavy metals in sediments in an urban river. Environ Sci Eur 33:42. https://doi.org/10.1186/s12302-021-00487-x
32. USEPA (2004) The incidence and severity of sediment contamination in surface waters of the United States, National Sediment Quality Survey. EPA 823-R-04-007, 2nd edn. US Environmental Protection Agency Office of Water, Washington, DC
33. USEPA (2005) Procedures for the derivation of equilibrium partitioning sediment benchmarks (ESBs) for the protection of benthic organisms: metal mixtures (Cadmium, Copper, Lead, Nickel, Silver and Zinc), Office of Research and Development, Washington, DC, EPA-600-R-02-011
34. Jiang M, Sheng Y, Liu Q, Wang W, Liu X (2021) Conversion mechanisms between organic sulfur and inorganic sulfur in surface sediments in coastal rivers. Sci Total Environ 752. https://doi.org/10.1016/j.scitotenv.2020.141829
35. Zhang Y, Yang J, Simpson SL, Wang Y, Zhu L (2019) Application of diffusive gradients in thin films (DGT) and simultaneously extracted metals (SEM) for evaluating bioavailability of metal contaminants in the sediments of Taihu Lake, China. Ecotoxicol Environ Saf 184:109627. https://doi.org/10.1016/j.ecoenv.2019.109627
36. Li D, Liu X, Zhanguang L, Zhao X (2016) Variations in total organic carbon and acid-volatile sulfide distribution in surface sediments from Luan River Estuary, China. Environ Earth Sci 75:1073. https://doi.org/10.1007/s12665-016-5873-1

37. Burton ED, Phillips IR, Hawker DW (2005) Reactive sulfide relationships with trace metal extractability in sediments from southern Moreton Bay, Australia. Mar Pollut Bull 50:589–608. https://doi.org/10.1016/j.marpolbul.2005.01.022
38. Europäische Gemeinschaft (EG), 2006/Draft2009. Verordnung (EG) Nr. 1907/2006 des Europäischen Parlaments und des Rates vom 18. Dezember 2006 zur Registrierung, Bewertung, Zulassungund Beschränkung chemischer Stoffe (REACH), zur Schaffungeiner Europäischen Agentur für chemische Stoffe, zur Änderung der Richtlinie 1999/45/EG und zur Aufhebung der Verordnung (EWG) Nr. 793/93 des Rates, der Verordnung (EG) Nr. 1488/94 der Kommission, der Richtlinie 76/769/EWG des Rates sowie der Richtlinien 91/155/EWG, 93/67/EWG, 93/105/EG und 2000/21/EG der Kommission. Amtsblatt der Europäischen Union, Nr. L 396/1. 851 S.
39. Ahlf W, Drost W, Heise S (2009) Incorporation of metal bioavailability into regulatory frameworks—metal exposure in water and sediment. J Soils Sediments 9:411–419. https://doi.org/10.1007/s11368-009-0109-6
40. Shine JP, Trapp CJ, Coull BA (2003) Use of receiver operating characateristic curves to evaluate sediment quality guidelines for metals. Environ Toxicol Chem 22(7):1642–1648. https://doi.org/10.1002/etc.5620220728
41. Ankley GT, Di Toro DM, Hansen DJ, Berry WJ (1996) Technical basis and proposal for deriving sediment quality criteria for metals. Environ Toxicol Chem 15:2056–2066. https://doi.org/10.1002/etc.5620151202
42. Lee JS, Lee BG, Yoo H, Koh CH, Luoma SN (2001) Influence of reactive sulfide (AVS) and supplementary food on Ag, Cd, and Zn bioaccumulation in the marine polychaete Neanthes arenaceodentata. Mar Ecol Prog Ser 216:129–140. https://doi.org/10.3354/meps216129
43. Lee JS, Lee BG, Luoma SN, Yoo H (2004) Importance of equilibration time in the partitioning and toxicity of zinc in spiked sediment bioassays. Environ Toxicol Chem 23:65–71. https://doi.org/10.3354/meps216129
44. Ankley GT, Phipps GL, Leonard EN, Benoit DA, Mattson VR, Kosian PA, Mahony JD (1991) Acid-volatile sulfide as a factor mediating cadmium and nickel bioavailability in contaminated sediments. Environ Toxicol Chem 10(10):1299–1307. https://doi.org/10.1002/etc.5620101009
45. Allen HE, Fu G, Deng B (1993) Analysis of acid-volatile sulfide (AVS) and simultaneously extracted metals (SEM) for the estimation of potential toxicity in aquatic sediments. Environ Toxicol Chem 12:1441–1453. https://doi.org/10.1002/etc.5620120812
46. Prica M, Dalmacija B, Rončević S, Krčmar D, Bečelić M (2008) A comparison of sediment quality results with acid volatile sulfide (AVS) and simultaneously extracted metals (SEM) ratio in Vojvodina (Serbia) sediments. Sci Total Environ 389(2–3):235–244. https://doi.org/10.1016/j.scitotenv.2007.09.006
47. Campana O, Blasco J, Simpson SL (2013) Demonstrating the appropriateness of developing Sediment quality guidelines based on sediment geochemical properties. Environ Sci Technol 47:7483–7489. https://doi.org/10.1021/es4009272
48. Vangheluwe MLU, Verdonck FAM, Besser JM, Brumbaugh WG, Ingersoll CG, Schlekat CE, Garman ER (2013) Improving sediment-quality guidelines for nickel:development and application of predictive bioavailability models to assess chronic toxicity of nickel in freshwater sediments. Environ Toxicol Chem 32:2507–2519. https://doi.org/10.1002/etc.2373
49. Costello DM, Hammerschmidt CR, Burton GA (2016) Nickel partitioning and toxicity in sediment during aging: variation in toxicity related to stability of metal partitioning. Environ Sci Technol 50:11337–11345. https://doi.org/10.1021/acs.est.6b04033
50. Besser JM, Ingersoll CG, Giesty JP (1996) Effects of spatial and temporal variation of acid-volatile sulfide on the bioavailability of copper and zinc in freshwater sediments. Environ Toxicol Chem 15(3):286–293. https://doi.org/10.1002/etc.5620150310
51. Besser JM, Brumbaugh WG, May TW, Ingersoll CG (2003) Effects of organic amendments on the toxicity and bioavailability of cadmium and copper in spiked formulated sediments. Environ Toxicol Chem 22:805–815. https://doi.org/10.1002/etc.5620220419
52. Strom D, Simpson SL, Batley GE, Jolley DF (2011) The influence of sediment particle size and organic carbon on toxicity of copper to benthic invertebrates in oxic/suboxic surface sediments. Environ Toxicol Chem 30:1599–1610. https://doi.org/10.1002/etc.531

53. Zhang T, Zhong W, Zeng Y, Zhu L (2012) Sediment heavy metals quality criteria for fresh waters basedon biological effect database approach. Chin J Appl Ecol 23:2587–2594. (in Chinese). PMID: 23286020
54. Zhang YF, Han YW, Yang JX, Zhu LY, Zhong WJ (2017) Toxicities and risk assessment of heavy metals in sediments of Taihu Lake, China, based on sediment quality guidelines. J Environ Sci 62:31–38. https://doi.org/10.1016/j.jes.2017.08.002
55. Casas AM, Crecelius EA (1994) Relationship between acid volatile sulfide and the toxicity of zinc, lead and copper in marine sediments. Environ Toxicol Chem 13(3):529–536. https://doi.org/10.1002/etc.5620130325
56. USEPA (1994) Methods for measuring the toxicity and bioaccumulation of sediment-associated contaminants with freshwater invertebrates. 600/R-94/024. Washington, DC
57. USEPA (1989) Briefing report to the EPA science advisory board on the equilibrium partitioning approach to generating sediment quality criteria. EPA 440/5-89-002. Office of Water Regulations and Standards, Criteria and Standards Division, Washington, DC
58. Di Toro DM, Zarba CS, Hansen DJ, Berry WJ, Swartz RC, Cowan CE, Pavlou SP, Allen HE, Thomas NA, Paquin PR (1991) Technical basis for establishing sediment quality criteria for nonionic organic chemicals using equilibrium partitioning. Environ Toxicol Chem 10(12):1541–1583. https://doi.org/10.1002/etc.5620101203
59. USEPA (2000) Office of Water of Science and Technology, Draft implementation framework for the use of equilibrium partitioning sediment quality guideline. Washington DC: 4217
60. Meng W, Zhang Y, Zheng BH (2006) The quality criteria, standards of water environment and the water pollutant control strategy on watershed. Res Environ Sci 19:1–6 (in Chinese with English abstract)
61. Van der Kooij LA, Van de Meent D, Van Leeuwen CJ, Bruggeman WA (1991) Deriving quality criteria for water and sediment from the results of aquatic toxicity tests and product standards: application of the equilibrium partitioning method. Water Res 25:697–705. https://doi.org/10.1016/0043-1354(91)90045-R
62. Webster J, Ridgway I (1994) The application of the equilibrium partitioning pproach for establishing sediment quality criteria at two UK sea disposal and outfall sites. Mar Pollut Bull 28:653–661. https://doi.org/10.1016/0025-326x(94)90300-x
63. Azzoni R, Giordani G, Viaroli P (2005) Iron–sulphur–phosphorus interactions: implications for sediment buffering capacity in a Mediterranean eutrophic lagoon (Sacca di Goro, Italy). Hydrobiologia 550(1):131–148. https://doi.org/10.1007/S10750-005-4369-X
64. Zhu MX, Liu J, Yang GP, Li T, Yang RJ (2012) Reactive iron and its buffering capacity towards dissolved sulfide in sediments of Jiaozhou Bay, China. Mar Environ Res 80:46–55. https://doi.org/10.1016/J.MARENVRES.2012.06.010
65. Ming J, Yanqing Sh, Qunqun L, Wenjing W, Xiaozhu L (2021) Conversion mechanisms between organic sulfur and inorganic sulfur in surface sediments in coastal rivers. Sci Total Environ 752:141829. https://doi.org/10.1016/j.scitotenv.2020.141829
66. Signorini A, Massini G, Migliore G, Tosoni M, Varrone C, Izzo G (2008) Sediment biogeochemical differences in two pristine Mediterranean coastal lagoons (in Italy) characterized by different phanerogam dominance-a comparative approach. Aquat Conserv Mar Freshwat Ecosyst 18(S1):S27–S44. https://doi.org/10.1002/AQC.953
67. Lefcheck JS (2016) Piecewise SEM: piecewise structural equation modelling in R for ecology, evolution, and systematics. Methods Ecol Evol 7:573–579. https://doi.org/10.1111/2041-210X.12512
68. Pellegrini E, Contin M, Livia VA, Chiara F, Nobili MD (2019) Soil organic carbon and carbonates are binding phases for simultaneously extractable metals (SEM) in calcareous saltmarsh soils. Environ Toxicol Chem 38:2688–2697. https://doi.org/10.1002/etc.4590
69. El Zokm GM, Okbah MA, Younis AM (2015) Assessment of heavy metals pollution using AVS-SEM and fractionation techniques in Edku Lagoon sediments, Mediterranean Sea. J Environ Sci Health Part A Toxic/Hazard Subst Environ Eng 50:571–584. https://doi.org/10.1080/10934529.2015.994945

70. Okbah M A, Younis AM, El Zokm GM (2015) Heavy metals fractionation and acid volatile sulfide (AVS) in the Bardawil lagoon sediments, northern Sinai, Egypt. Dev Analyt Chem 2. https://doi.org/10.14355/dac.2015.02.001
71. Li Z, Ma T, Sheng Y (2022) Ecological risks assessment of sulfur and heavy metals in sediments in a historic mariculture environment, North Yellow Sea. Mar Pollut Bull 183:114083. https://doi.org/10.1016/j.marpolbul.2022.114083
72. Gao X, Song J, Li X et al (2020) Sediment quality of the Bohai Sea and the northern Yellow Sea indicated by the results of acid-volatile sulfide and simultaneously extracted metals determinations. Mar Pollut Bull 155(3):111147. https://doi.org/10.1016/j.marpolbul.2020.111147
73. Ju Y-R, Chen C-F, Lim Y, Tsai C-Y, Chen Ch W, Dong Ch D (2022) Developing ecological risk assessment of metals released from sediment based on sediment quality guidelines linking with the properties: a case study for Kaohsiung Harbor. Sci Total Environ 852:158407. https://doi.org/10.1016/j.scitotenv.2022.158407
74. Macdonald DD, Ingersoll CG, Berger TA (2000) Development and evaluation of consensus-based sediment quality guidelines for freshwater ecosystems. Arch Environ Contam Toxicol 39:20–31. https://doi.org/10.1007/s002440010075
75. Ingersoll CG, MacDonald DD, Wang N et al (2001) Predictions of sediment toxicity using consensus-based freshwater sediment quality guidelines. Arch Environ Contam Toxicol 41:8–21. https://doi.org/10.1007/s002440010216
76. Leonard EN, Cotter AM, Ankley GT (1996) Modified diffusion method for analysis of acid volatile sulfides and simultaneously extracted metals in freshwater sediment. Environ Toxicol Chem 15(9):1479–1481. https://doi.org/10.1002/etc.5620150908
77. Hübner R, Astin KB, Herbert RJH (2009) Comparison of sediment quality guidelines (SQGs) for the assessment of metal contamination in marine and estuarine environments. J Environ Monit 11:713–722. https://doi.org/10.1039/b818593j
78. Burton GA, Green A, Baudo R, Forbes V, Nguyen LT, Janssen CR, Kukkonen J, Leppanen M, Maltby L, Soares A, Kapo K, Smith P, Dunning J (2007) Characterizing sediment acid volatile sulfide concentrations in European streams. Environ Toxicol Chem 26(1):1–12. https://doi.org/10.1897/05-708r.1
79. Chaikaew P, Sompongchaiyakul P (2018) Acid volatile sulphide estimation using spatial sediment covariates in the Eastern Upper Gulf of Thailand: multiple geostatistical approaches. Oceanologia 60(4):478–487. https://doi.org/10.1016/j.oceano.2018.03.003
80. Campana O, Rodriguez A, Blasco J (2009) Identification of a potential toxic hot spot associated with AVS spatial and seasonal variation. Arch Environ Contam Toxicol 56:416–425. https://doi.org/10.1007/s00244-008-9206-6
81. Poot A, Meerman E, Gillissen F, Koelmans AA (2009) A kinetic approach to evaluate the association of acid volatile sulfide and simultaneously extracted metals in aquatic sediments. Environ Toxicol Chem 28(4):711–717. https://doi.org/10.1897/08-506.1. PMID: 19007305
82. Ramanathan RS, Nasr-El-Din HA, Zakaria AS (2020) New insights into the dissolution of iron sulfide using chelating agents. SPE J 25:3145–3159. https://doi.org/10.2118/202469-PA
83. Di Toro DM, Mahony JD, Gonzalez AM (1996) Particle oxidation model of synthetic FeS and sediment acid-volatile sulfide. Environ Toxicol Chem 15:2156–2167. https://doi.org/10.1002/etc.5620151211
84. Landers DH, David MB, Mitchell MJ (1983) Analysis of organic and inorganic sulfur constituents in sediments, soils and water. Int J Environ Anal Chem 14:245–256. https://doi.org/10.1080/03067318308071623
85. Hinkey LM, Zaidi BR (2007) Differences in SEM–AVS and ERM–ERL predictions of sediment impact from metals in two US Virgin Islands marinas. Mar Pollut Bull 54(2):180–185. https://doi.org/10.1016/j.marpolbul.2006.09
86. Rickard D, Morse JW (2005) Acid volatile sulfide (AVS). Mar Chem 97(3–4):141–197. https://doi.org/10.1016/j.marchem.2005.08.004
87. Tisserand D, Guédron S, Razimbaud S, Findling N, Charlet L (2021) Acid volatile sulfides and simultaneously extracted metals: a new miniaturized 'purge and trap' system for laboratory and field measurements. Talanta 233:22490. https://doi.org/10.1016/j.talanta.2021.122490

88. Hernández-Crespo C, Martín M, Ferrís M, Oñate M (2012) Measurement of acid volatile sulphide and simultaneously extracted metals in sediment from Lake Albufera (Valencia, Spain). Soil Sediment Contamination Int J 21(2):176–191. https://doi.org/10.1080/15320383.2012.649374
89. Ngiam LS, Lim PE (2001) Speciation patterns of heavy metals in tropical estuarine anoxic and oxidized sediments by different sequential extraction schemes. Sci Total Environ 275:53–61. https://doi.org/10.1016/s0048-9697(00)00853-6
90. Chai M, Shen X, Li R, Qiu G (2015) The risk assessment of heavy metals in Futian mangrove forest sediment in Shenzhen Bay (South China) based on SEM-AVS analysis. Mar Pollut Bull 15, 97(1–2):431–439. https://doi.org/10.1016/j.marpolbul.2015.05.057
91. Spencer KL, Dewhurst RE, Penna P (2006) Potential impacts of water injection dredging on water quality and ecotoxicity in Limehouse Basin, River Thames, SE England, UK. Chemosphere 3:509–521. https://doi.org/10.1016/j.chemosphere.2005.08.009
92. Li F, Zeng XY, Yu YJ, Wu CH, Mai G, Song WW, Wen YM, Duan ZP, Yang JY (2014) A field study of the relationship between sulfide-bound metals and bioaccumulation by *Limnodrilus sp.* in a heavily polluted river. Environ Monitor Assessment 186(8):4935–46. https://doi.org/10.1007/s10661-014-3749-y. Epub 2014 Apr 4. PMID: 24700206
93. Van Griethuysen C, Meijboom EW, Koelmans AA (2003) Spatial variation of metals and acid volatile sulfide (AVS). in floodplain lake sediment. Environ Toxicol Chem 22(3):457-465. https://doi.org/10.1007/s10661-014-3749-y

Chapter 5
A Review and an Outlook "AVS-SEM" Studies in Egypt Compared to Aquatic Environment Around the World

Rather than estimating "total metals", the AVS/SEM technique provides a more precise picture of sediment toxicity. However, only a few research studies have utilized AVS and SEM to evaluate the state of the aquatic environment. Because AVS in sediment is highly dynamic, this could explain why the results of the AVS–SEM models differed significantly at different times. Furthermore, AVS–SEM measurements are sensitive to environmental conditions. A deep insight into the contribution of each extracted metal to ΣSEM from different aquatic environments revealed in most investigations that SEM (Zn) has a preponderance over all other metals, with a proportion of 41–97% of ΣSEM. The far more hazardous Cd, on the other hand, contributed less than 1% to ΣSEM (Table 5.1; constructed by the author). The solubility of metal sulfides differed, which influenced the percentage of various metals in ΣSEM [1–8].

Studies on acid-volatile sulfide topics in the Egyptian aquatic environment are rare. However, [1, 2, 4, 5] assessed the Egyptian Mediterranean coast and Northern Egyptian lakes along the Mediterranean through the application of AVS-SEM models. This review showed an in-depth look at the research done in Egypt and established the baseline for future research.

Naser et al. (1), studied AVS-SEM correlations and potential bioavailability of heavy metals in marine sediments as the first effort on the Egyptian Mediterranean coast. The distribution of AVS and SEM (Zn, Cu, Ni, Cd, and Pb) was measured in 20 sample locations along the Egyptian Mediterranean coast. The values measured by SEM varied from 0.012 to 0.241 μmol g^{1-}. The quantities of AVS varied greatly over the research region, ranging from 0.015 to 31.326 μmol g^{1-}. To assess potential bioavailability, three AVS-SEM models were used: SEM–AVS, SEM/AVS, and SEM-AVS/foc.

Sediments at the Eastern Harbour, Western Harbour, Ras El-Burr, El-Gamil East, and Port Said are potentially hazardous in the context of the SEM/AVS concept, but no indication of related detrimental toxic impact would arise based on the SEM-AVS approach. $\sum$SEM-AVS]/$f$oc was < 130 μmol g^{-1} in all sediment samples revealing a low threat of adverse biological. Another approach using Probable Effect

G. M. El Zokm, *Ecological Quality Status of Marine Environment*, Earth and Environmental Sciences Library, https://doi.org/10.1007/978-3-031-29203-3_5

Table 5.1 AVS and SEM research in Egypt compared to similar investigations across the world

Locations			AVS	$\sum$SEM (μmol/g)	SEM_{Zn}	SEM_{Cu}	SEM_{Pb}	SEM_{Ni}	SEM_{Cd}	References
			(μmol/g)		(μmol/g)	(μmol/g)	(μmol/g)	(nmol/g)	(μmol/g)	
Edku lagoon Egypt		Mean	10.4	2.612	1.929	0.292	0.156	0.247	0.008	[5]
BaradawIll lagoon Egypt		Mean	0.627	0.598	1.296	0.142	0.059	0.099	0.012	[4]
Western region in Egypt along Mediterranean		Mean	3.307	0.128	0.078	0.021	0.016	0.012	0.0004	[2]
Manzala lagoon Egypt		Mean	26.128	2.894	2.089	0.411	0.175	0.211	0.008	[1]
Bohai Sea China		Range	0.39–3.99	0.54–1.46	0.289–0.782	0.091–0.308	0.037–0.099	0.096–0.273	0.0002–0.0017	[8]
		Mean	1.25	1.05	0.572	0.211	0.071	0.192	0.0007	
Bizerte Lagoon (Tunisia)		Mean	0.7	1.063	0.851	0.091	0.07	0.051		[7]
		Range	0.017–22.74	0.22–20.15	0.13–12.43	0.09–6.34	0.009–1.14	0.05–0.034	0.0005–0.098	[6]
Asaluyeh harbor, Iran		Mean	4.63	7.72	4.44	1.86	0.39	0.22	0.0026	
Futian mangrove, Shenzhen Bay		Mean	2.93	2.37	2.06	0.130	0.17	ND	0.0113	[11]
Leizhou Peninsula, China		Range	0.11–55.55	0.03–8.60	0.02–8.21	0.001–0.161	0.001–0.185	0.001–0.069	0.03–0.826	[12]
		Mean	4.45	0.84	0.76	0.031	0.045	0.01	0.146	
Laizhou Bay, Bohai Sea	Summer	Mean	4.98	0.48	0.206	0.104	0.029	0.122	0.0008	[9]
	Autumn	Mean	3.61	0.35	0.163	0.072	0.023	0.089	0.0004	
River connected to Laizhou Bay	Summer	Mean	26.96	1.07	0.702	0.218	0.041	0.106	0.0007	
	Autumn	Mean	21.83	6.64	6.35	0.158	0.033	0.099	0.0069	
Zhangzi Island, China		Range	0.71–11.03	0.10–0.57	0.052–0.216	0.026–0.247	0.01–0.067	0.013–0.046	0–1.3	[13]

(continued)

Table 5.1 (continued)

Locations		AVS (μmol/g)	$\sum$SEM (μmol/g)	SEM_{Zn} (μmol/g)	SEM_{Cu} (μmol/g)	SEM_{Pb} (μmol/g)	SEM_{Ni} (nmol/g)	SEM_{Cd} (μmol/g)	References
	Mean	4.05	0.32	0.137	0.109	0.039	0.034	0.7	
Laizhou Bay, China	Range	1.22–7.60	0.20–0.74	0.064–0.278	0.051–0.310	0.014–0.097	0.033–0.138	1.1–3.3	[13]
	Mean	2.99	0.45	0.165	0.166	0.044	0.074	2	
Zhangjiang Estuary, Fujian,	Range	0.2–12.5	1.4–2.1	0.731–1.391	0.152–0.324	0.115–0.197	0.095–0.207	0.623–1.690	[14]
Vojvodina, Serbia	Range	3.1–14.85	3.71–14.85	1.81–10.98	0.10–1.52	0.01–0.98	0.41–3.84	0.01–0.21	[10]
	Mean	8.33	8.45	5.33	0.53	0.44	1.11	0.07	

Levels (PEL) was utilized to predict toxicity through mean PEL quotients (PELq = $\_\sum$ [Ci/(PELi)]/n). All sample locations had a mean PELq range of (0.11–1.5) and were categorized as toxic, except Nobarreya, Baghoush, and the Western Harbour with PELq < 0.1 and were rated nontoxic. This research concluded that trace metal mobility and bioavailability in surface sediments from the Egyptian Mediterranean coastal area are poor. The results of this investigation are used as baselines for acid-volatile sulfide and simultaneously extracted metals sediments throughout the Egyptian coast along the Mediterranean. The findings of this investigation revealed significant variations in AVS levels in sediments along the Egyptian Mediterranean from El-Salloum to Rafah but still low levels might be due to the sediments' low organic matter concentration. It's worth noting that the SEM/AVS, SEMAVS/f_{oc} findings were accordant with the mean PEL quotients evaluated for the sample locations. The findings revealed that trace metals in sediments from the Egyptian Mediterranean coast have limited remobility and bioavailability.

Okbah et al. [4], used heavy metal fractionation and AVS-SEM models to describe the apportionment of iron, copper, lead, and cadmium in Bardawil Lagoon sediments and assess sediment quality. Surface sediments collected from ten sites in Bardawil Lagoon were analyzed for grain size, $CaCO_3$, and organic carbon content. The proportion of $CaCO_3$ in the findings ranged from 53.5 to 70.5%. The $\sum$SEM/AVS ratio in Bardwell lagoon sediments was higher than one ($\sum$SEM/AVS varied from 1.28 to 9.19). Sediments would be identified as hazardous based on the information from this study. Using a sequential extraction technique, the exchangeable amounts of Cu, Fe, Pb, and Cd in the sediments were examined. The results of lead and cadmium fractionation revealed the hazards of these elements, with non-residual fractions accounting for more than 75% of the total. In the research region, trace metals (Cu, Pb, and Cd) make up a large portion of the labile form, accounting for more than 60% of the overall concentration. For the research area, the Risk Assessment Code (RAC) for Cu, Pb, and Cd concluded medium to high risk. However, iron indicated nil to low risk.

This review included a rapid screening for several studies around the world that used AVS-SEM models, as shown in Table 5.1.

Gao et al. [8], assessed the quality of sediments from the northern Yellow Sea in China and the Bohai Sea using AVS and SEM. Based on [SEM] and [AVS] differences, in the Bohai Sea, more than 60% of the locations were classified to have probable adverse impacts on aquatic life, whereas a comparable percentage in the northern Yellow Sea was 25%. The particle size, TOC, pH, and water content all had a significant impact on the distribution of [SEM] and the [SEM]/[AVS] ratios; however, the distribution of [AVS] was primarily governed by the sediment's redox state. Furthermore, when organic carbon is taken into account in this study; ([SEM]-[AVS])/fOC 130 mol g^{1-} OC, indicating that there were no negative impacts.

Arfaeinia et al. [6], reported an obvious variation in SEM-AVS differences between sediments from urban areas and industrial areas in Asaluyeh harbor. Since AVS and SEM concentrations have varied due to human activities. In this study, the AVS was shown to be positively associated with the SEM. during seasons, indicating that SEM was carried by this key carrier. There was no substantial difference in AVS levels in spring and autumn. SEM levels in urban sediments, on the other hand, revealed significant variations. Based on (SEM-AVS)/*f*OC and SEM-AVS approaches, 20% of stations in autumn and 47% of stations in spring have no adverse effect.

Zaaboub et al. [7], investigated the Bizerte Lagoon, southern Mediterranean, to assess sediment quality against SEM-AVS. Environmental disturbance at the water–sediment interface is utilized as a biogeochemical tool to drive inferences on pollution status using SEM and AVS. SEM–AVS/*f*OC model in sediment verified the possible bioavailability of metals, especially Zn. The authors certified the critical rule of the bioavailable fraction of SEM_{Metal} on sediment toxicity.

Zhuang and Gao [9], collected samples from Laizhou Bay and the rivers that connect it during summer and autumn. Industrial pollution had a greater impact on AVS and SEM in riverine sediments than in marine sediments. AVS-SEM was applied to predict sediment quality. Accordingly, by using the $[\sum SEM–AVS]/f_{OC}$ model, only the surface sediment at one location in autumn has a negative biological effect. According to the AVS–$\sum$SEM criterion, all other site sediments may have no adverse impact in both seasons. According to the t-test, seasonal fluctuations had a significant impact on the apportionment of SEM in marine sediments, but not in riverine sediments. SEM and AVS development and existence in riverine sediments are more complex than in marine sediments.

Prica et al. [10], explored total metals, pore-water metal levels, and AVS-SEM method in Vojvodina (Serbia) sediments. The AVS varied from 3.10 to 15.97 mg kg^{-1}. AVS levels above 10 μmol g^{-1} are found in around 40% of the samples currently being studied. Seasonal changes and the dynamic behaviour of the underlying water stream might both contribute to the AVS variance. SEM levels tend to fluctuate throughout the year, ranging from 3.71 μmol g^{-1} to 14.85 μmol g^{-1}. Despite the fact that the associated SEM/AVS ratio ranged from 0.4 to 3.3, it was less than 1 in 65% of the samples. Although AVS and SEM concentrations in sediments fluctuated throughout the year, the SEM/AVS ratio followed a similar pattern. According to the Netherlands' interim sediment quality guideline (SQG), 48% of all examined sediment samples were severely contaminated.

5.1 Case Study 1 (Egypt; Variation of AVS/SEM in Coastal Lakes)

Younis et al. [1] assess the heavy metal contamination in sediments from three northern delta lagoons (Maryut, Burullus, and Manzalah) that connect to the Mediterranean Sea in Egypt using an AVS/SEM approach as the first approach. The heavy metals being studied are zinc, iron, copper, lead, nickel, and cadmium.

5.1.1 Sampling

Fourty three (43) sediment samples were taken from three lagoons and their drains (at a depth of 20 cm). In the winter of 2011, duplicate sediment samples were obtained using a stainless-steel grab sampler from Hydro-Bios and maintained frozen at 20 °C waiting for examination (USEPA 2000).

5.1.2 Methodology

Grain size, carbonate, and total organic carbon were measured according to [15–17] respectively. Electrical conductivity, pH, dissolved oxygen, chemical oxygen demand, biological oxygen, oxidizable organic matter, and total heavy metals in sediment are considered followed [18–20], respectively. The purge and trap technique was used to measure AVS-SEM [21].

5.1.3 Results and Discussions

The findings of this investigation screened the regional variance of AVS and SEM sediment samples from 43 places in Manzalah, Burullus, and Maryut lagoons as predictors of the bioavailability of several divalent metals (Cu, Zn, Cd, Pb, and Ni) in sediments as well as signs of metal toxicity in these sediments as a first report. According to the findings, sediments from Burullus lagoon exhibited greater amounts of $\sum$SEM (Cu + Zn + Cd + Pb + Ni) values than Maryut's and Manzalah's. AVS concentrations, on the other hand, were noticeably greater in the lagoons Manzalah and Maryut and appeared to be compatible with the rise in organic matter compared to the lagoon Burullus. The average SEM concentrations in all lagoons were in the order Zn > Cu > Ni > Pb > Cd. All sampling locations, with the exception of four stations in lagoon Burullus and one station in lagoon Maryut, had SEM/AVS ratios that were less than 1. (SEM/AVS > 1), which implies that the metals in these sediments have the potential to be hazardous.

5.1.4 Statistical Analysis

Table 5.2 displays the correlation between the research area's AVS, SEM-Zn, SEM-Cu, SEM-Cd, SEM-Pb, SEM-Ni, total organic matter (TOM), sand, silt, and clay content, and water content. The findings indicated a substantial positive correlation between SEM-Cu, SEM-Ni, SEM-Pb, and SEM-Cd and SEM-Zn with coefficients of 0.80, 0.94, 0.84, and 0.55 (confidence level = 0.01). This shows that the solubility of

Table 5.2 Analysis of the relationships between SEM, AVS, water content, TOM, sand, salt, and clay [1]

	SEM-Cu	SEM-Zn	SEM-Cd	SEM-Pb	SEM-Pb	AVS	Water content	TOM	Sand	Silt + clay
SEM-Cu	1				SEM-Ni					
SEM-Zn	0.806	1								
SEM-Cd	0.552	0.817	1							
SEM-Pb	0.848	0.826	0.596	1						
SEM-Ni	0.942	0.912	0.656	0.901	1					
AVS	0.005	− 0.0115	− 0.056	− 0.013	− 0.066	1				
Water content	0.481	0.617	0.571	0.464	0.540	− 0.054	1			
TOM	0.408	0.473	0.452	0.489	0.398	0.261	0.545	1		
Sand	− 0.0231	− 0.0395	− 0.300	− 0.162	− 0.324	0.077	− 0.667	− 0.298	1	
Silt + clay	0.0231	0.0395	0.300	0.162	0.324	− 0.077	0.667	0.298	− 1.000	1

CuS, NiS, PbS, and CdS has a significant impact on the solubility of ZnS. SEM-Cu, SEM-Zn, SEM-Cd, SEM-Pb, and SEM-Ni had coefficients of 0.40, 0.47, 0.45, 0.48, and 0.39 (confidence level = 0.01) that were positively and substantially associated with organic matter. While SEM-Zn and SEM-Ni showed positive correlations to clay with lower coefficient values (0.39 and 0.32; confidence level = 0.05), SEM-Zn and SEM-Ni exhibited negative correlations to sand with coefficients of 0.39 and 0.32, respectively. According to these findings, organic matter and clay play a significant role in the formation of SEM in sediments under anoxic conditions in aquatic environments.

5.1.5 *Conclusion*

This research established a baseline for AVS-SEM in Egypt's lagoons (Maryut, Burullus, and Manzalah) as ecological indicators of divalent metal bioavailability in sediments. The SEM–AVS pattern in Burullus lagoon displays a ratio of < 1 in all places except two drains; SEM/AVS values varied from 0.10 to 2.56, whereas it ranged from 0.02 to 0.74 for lagoon Manzalah and from 0.01 to 1.50 for Maryut lagoon. Harmful impacts are not expected in the majority of sediments in the study area according to a USEPA assessment. This study reported that SEM–AVS models are more powerful ecological markers of metal availability than total concentrations of heavy metals (Fig. 5.1).

5.2 Case Study 2 (China; AVS-SEM/PEL-TEL)

Yin et al. [22], performed research to (1) describe the apportionment of the heavy metals in Lake Chaohu sediments, (2) assess the sediment quality in Lake Chaohu using AVS-SEM models and PEL-TEL values.

5.2.1 *Sampling*

Surface sediments (0–10 cm) from 27 sites across Lake Chaohu in China were collected in August 2008. Samples were put in polyethylene bags and stored on ice in the fridge before being arrived to the lab and kept under nitrogen at temperatures below 4 °C. The subsamples were then freeze-dried, crushed, and sieved through 0.063-mm mesh sieves before being kept at 4 °C until analysis.

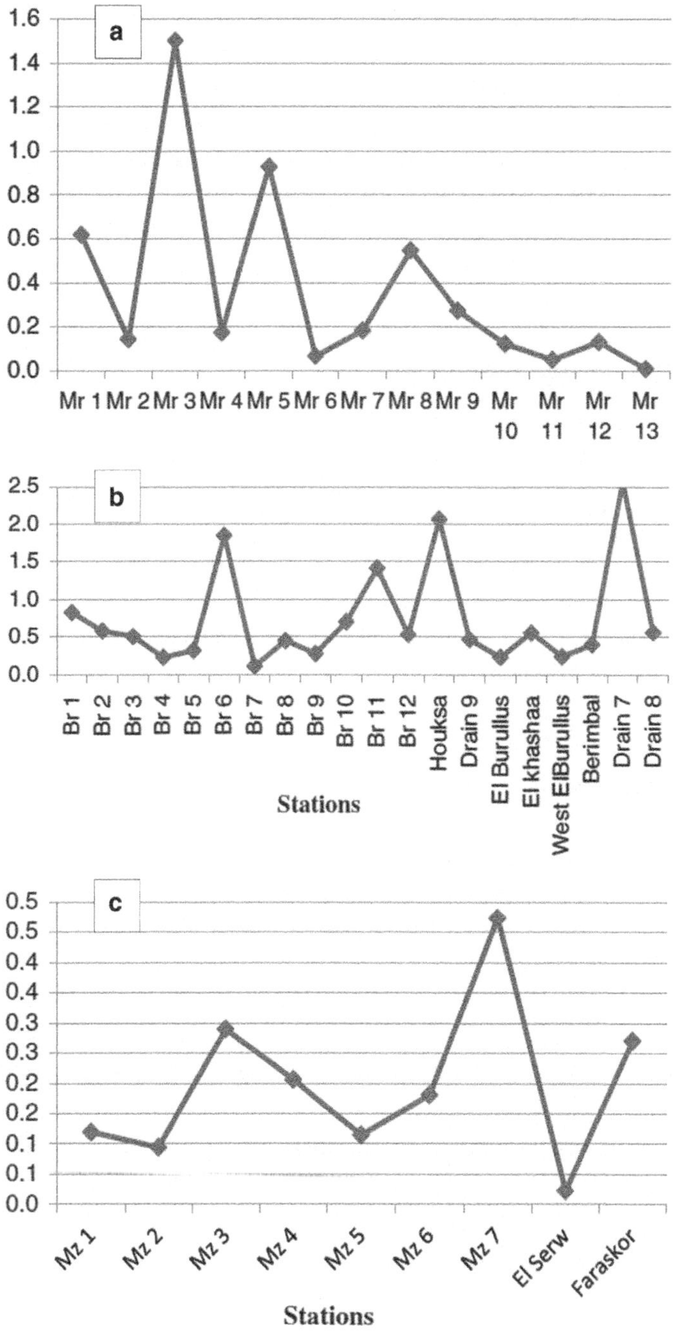

Fig. 5.1 Variations of SEM/AVS in sediments of **a** Maryut, **b** Brullus and **c** Manzala 2011 [1]

5.2.2 Methodology

In triplicate, AVS and SEM were recovered from sediment samples that had been maintained under nitrogen for no more than 48 h at temperatures below 4 °C. To prevent Fe (III) interference, AVS was measured with ascorbic acid using Hsieh's cold diffusion method [23]. The total heavy metals were evaluated using a microwave-assisted acid digestion technique (USEPA 1994). ICP-AES is used to measure total heavy metals and SEM.

5.2.3 Results and Discussions

Heavy metal average concentrations in the sediment samples were 49.82 mg/kg for Pb, 153.68 mg/kg for Zn, 26.23 mg/kg for Cu, 33.08 mg/kg for Ni and 0.43 mg/kg for Cd. Table 5.3 shows the sediment toxicity assessment findings derived using the AVS–SEM models, as well as TEL–PEL values. Based on the PEL values, 14 from 27 sample locations were expected to be toxic to benthic biota. The AVS–SEM mechanical models produced nearly entirely different findings. The conflicting findings produced by the TEL–PEL, and AVS–SEM techniques might be attributable to the limitations of these SQGs.

5.2.4 Conclusion

TEL–PEL values were consistent with SEM–AVS, SEM/AVS, and SEM–AVS/foc findings in only eight, five, and five of the 27 sample sites, respectively. As a consequence, more research should be done, such as biological toxicity testing and metal binding phase analysis.

5.3 Case Study 3 (China; Antarctic Lake Core Sediments)

Chen et al. [24], examined the effects of sulfate reduction on the bioavailability and toxicity of trace metals by analyzing AVS-SEMs, trace metals Cu, Cd, and Zn, and their chemical speciation based on BCR sequential extraction technique considering depth parameter.

Table 5.3 Sediments toxicity evaluation calculated from SEM–AVS models and TEL–PEL values [22]

Site	TOC	AVS μmol/g	∑SEM μmol/g	∑SEM/AVS	∑SEM-AVS μmol/g	SEM-AVS/foc μmol/gOC	Metals exceeding TEL	Metals exceeding PEL
1	1.76	6.22	6.52	1.05	0.3	16.8	Pb, Zn, Ni	Zn
2	1.49	1.24	3.16	2.55	1.92	128.9	Pb, Zn, Ni	–
3	1.04	0.47	0.66	1.42	0.2	19	–	–
4	1.73	5.69	3.52	0.62	− 2.17	− 125.5	Cu, Pb, Zn, Cd, Ni	Ni
5	1.15	0.21	0.63	3.01	0.42	36.70	Ni	–
6	1.54	5.32	2.61	0.49	− 2.71	− 176.60	Pb, Zn, Cd, Ni	Ni
7	1.27	2.36	4.31	1.83	1.96	153.90	Pb, Ni	–
8	1.61	1.12	7.70	6.85	6.58	408.20	Cu, Pb, Zn, Cd, Ni	Ni
9	1.61	2.92	5.01	1.72	2.09	129.60	Cu, Pb, Zn, Cd, Ni	Zn, Ni
10	1.59	1.79	4.43	2.47	2.63	165.20	Pb, Zn, Cd, Ni	–
11	1.51	5.18	4.22	0.81	− 0.96	− 63.70	Cu, Pb, Zn, Cd, Ni	Ni
12	1.43	0.99	6.14	6.21	5.15	361.50	Pb, Zn, Ni	Ni
13	1.47	0.77	0.84	1.09	0.07	4.91	Pb, Ni	–
14	1.59	2.56	2.23	0.87	− 0.32	− 20.3	Pb, Zn, Ni	Ni
15	1.40	1.95	0.85	0.44	− 1.09	− 74.20	Pb, Ni	–
16	1.09	1.81	0.96	0.53	− 0.85	− 77.9	Pb, Ni	–
17	1.20	1.71	1.44	0.84	− 0.27	− 22.4	Pb, Ni	–
18	1.36	1.88	1.48	0.79	− 0.40	− 29.2	Pb, Ni	–
19	1.20	0.54	1.18	2.20	0.64	53.70	Pb, Ni	–
20	1.28	1.37	1.26	0.92	− 0.10	− 8.16	Pb, Ni	Ni
21	1.51	0.84	1.24	1.48	0.40	26.60	Pb, Ni	Ni
22	1.09	0.21	1.41	6.71	1.20	110.50	–	–
23	1.67	1.84	1.40	0.76	− 0.44	− 26.1	Pb, Ni	Ni
24	1.42	1.24	1.23	0.99	− 0.01	− 0.75	Pb, Ni	Ni
25	1.53	1.26	1.52	1.21	0.26	17.10	Pb, Ni	
26	0.99	0.01	0.47	60.90	0.46	46.70	–	
27	1.74	3.29	1.42	70.43	− 1.87	− 107.8	Pb, Ni	Ni

5.3.1 Sampling

Two sediment cores, designated Y2-1 and YO, were taken from Yanou Lake on the Fildes Peninsula and Y2 lake on Ardley Island, respectively.

5.3.2 Methodology

$HClO_4$–HNO_3-HF were used for metal digestion. A modified version of the BCR-sequential extraction approach was utilized for the chemical speciation. The total phosphorus (TP), total nitrogen (TN), total organic carbon (TOC) and AVS were previously measured by [24]. One portion of the extracts was examined by ICP-MS for the presence of concurrently extracted metals (SEM) during the AVS analysis. Cu, Zn, and Cd were examined by (ICP-MS). To validate the BCR-sequential and simultaneous extraction, duplicate analysis was done.

5.3.3 Results and Discussions

The vertical distribution pattern of AVS was compatible with that of the oxidizable portion of Cu, Zn, and Cd in Y2-1 sediments (Figs. 5.2 and 5.3). But YO did not exhibit this reliability. The oxidizable fraction of trace metals consists of two components: one is the fraction that is bound to organic matter, and the other is the form of sulfides (AVS), indicating that the sulfide was the primary form of the oxidizable fraction and that S^{2-} from sulphate reduction in Y2-1 could efficiently combine ions of Cu and Cd to form AVS. The primary chemical speciation of Cu and Cd in the Y2-1 sediments was the oxidizable fraction, while Zn only made approximately 19.20% of the overall amount. This might be because trace metals vulcanize differently. For instance, S^{2-} could vulcanize trace metals like Ni and Cd largely (100 and 81%) but Zn could only be vulcanized by 11–16% [25]. These sulfide phases are especially susceptible to early dissolution during the sequential extraction procedure [26]. However, it is challenging to determine the scope of these issues without an independent investigation of metal speciation carried out using a different technique.

5.4 Case Study 4 (Egypt; Seasonal Variation of AVS-SEM/Fractionation)

For enrichment information, three integrated approaches: AVS-SEM, fractionation, and ISQG (Interim Sediment Quality Guidelines) to assess metal pollution in Edku

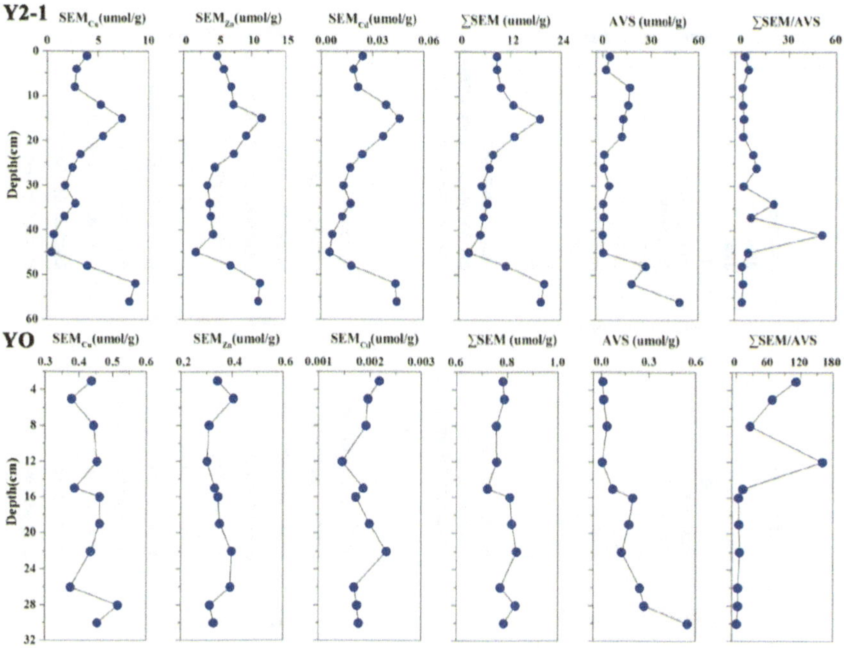

Fig. 5.2 Vertical apportionment of SEMs, AVS and ΣSEM/AVS in Y2-1 and YO sediments [24]

lagoon by [5] were reported, the heavy metals being studied are zinc, iron, copper, lead, nickel, and cadmium. As a first record seasonal variation is considered (summer and winter) in Edku Lagoon sediments.

5.4.1 Sampling

Sediments were collected from Edku lagoon (Brackish) at 13 places (at a depth of 20 cm). Duplicate sediment samples were collected (9 samples inside Edku lagoon and four samples from collected drains) (Barsik-El-kaairy-Edku-Bosily). To investigate the impact of seasonal fluctuation on the assessment methodologies, sediment samples were collected in two seasons (August 2010 and February 2011).

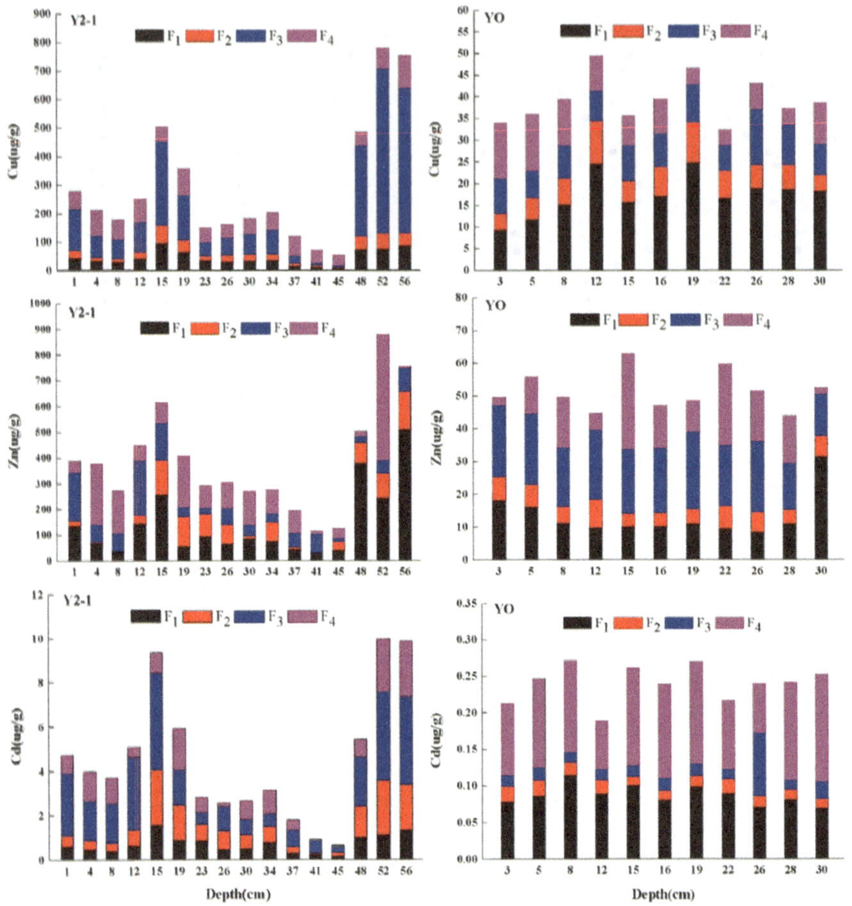

Fig. 5.3 Chemical speciation of Cu, Zn and Cd in Y2-1 and YO sediments [24]

5.4.2 Methodology

The same parameters as in case study 1 [1] were measured. In addition, sequential extraction was performed for studying metal fractionation by [29] and [30]. Also, ISQG are used for assessment heavy metals [27].

5.4.3 Statistical Analysis

A correlation matrix is used at a 95% confidence level and a p value of 0.05 is used. It was conducted on the dataset of AVS, SEM-Zn, SEM-Fe, SEM-Cu, SEM-Ni,

SEM-Cd, Sand, SEM-Pb, pH, silt and clay, salinity, organic matter OM, Conductivity, Temperature, dissolved oxygen (DO), chemical oxygen demand (COD), and biological oxygen demand (BOD) in Edku Lagoon to describe the behaviour and interaction between SEM with the other parameters. AVS, pH, and temperature have significant positive relationships, indicating the importance of temperature and pH in the chemistry of sulfide in the aquatic environment. In summer, the pH rises, as does the rate of bacterial sulfate reduction, lowering the concentration SEM. This causes a considerable rise in the proportion of immobilized sulfide minerals by lowering the concentration of dissolved sulfate [28]. A significant correlation between SEM-Fe and SEM-Cu revealed that the solubility of CuS was substantially impacted by the solubility of FeS, SEM-Zn, and SEM-Cd, all of which followed the same solubility trend. The powerful beneficial connection between OM and both SEM-Zn and SEM-Pb indicated that a large proportion of OM might be attributed to the breakdown of living creatures or come from the same anthropogenic source, which would increase the production of the two metals. SEM-Cd and COD had a significant negative relationship.

5.4.4 Conclusions

This study keeps an eye on the AVS-SEM systems in Edku lagoon as a first approach. Overall, the data showed that metals are retained in sediments through the production of metal sulfides, and that this capacity is far from exhausted. The spatial and temporal dependency of AVS-$\sum SEM_5$ was observed. Based on heavy metal fractionation patterns, organic materials, as well as Fe and Mn oxides, have been identified as the primary extra binding phases associated with heavy metals. Due to the findings in this investigation, a baseline for chemical-binding forms of heavy metals Fe, Zn, Cu, Pb, Cd, Ni, and AVS in Edku lagoon was created. Finally, metal mobility is found to be minimal in the Edku lagoon.

References

1. Younis AM, El Zokm GM, Okbah MA (2014) Spatial variation of acid-volatile sulfide and simultaneously extracted metals in Egyptian Mediterranean Sea lagoon sediments. Environ Monıt Assess 186:3567–3579. https://doi.org/10.1007/s10661-014-3639-3
2. Nasr SM, Khairy MA, Okbah MA, Soliman NF (2014) AVS-SEM relationships and potential bioavailability of trace metals in sediments from the Southeastern Mediterranean Sea, Eqypt. Chemi Ecol 30(1):15–28. https://doi.org/10.1080/02757540.2013.831080
3. He Y, Meng W, Xu J et al (2015) Spatial distribution and toxicity assessment of heavy metals in sediments of Liaohe River, northeast China. Environ Sci Pollut Res 22:14960–14970. https://doi.org/10.1007/s11356-015-4632-2
4. Okbah MA, Younis AM, El Zokm GM (2015) Heavy metals fractionation and acid volatile sulfide (AVS) in the Bardawil lagoon sediments, northern Sinai, Egypt. Dev Analyt Chem 2. https://doi.org/10.14355/dac.2015.02.001

5. El Zokm GM, Okbah MA, Younis AM (2015) Assessment of heavy metals pollution using AVS-SEM and fractionation techniques in Edku Lagoon sediments, Mediterranean Sea. J Environ Sci Health Part A Toxic/Hazard Subst Environ Eng 50:571–584. https://doi.org/10.1080/10934529.2015.994945
6. Arfaeinia H, Nabipour I, Ostovar A, Asadgol Z, Abuee E, Keshtkar M, Dobaradaran S (2016) Assessment of sediment quality based on acid-volatile sulfide and simultaneously extracted metals in a heavily industrialized area of Asaluyeh, Persian Gulf: concentrations, spatial distributions, and sediment bioavailability/toxicity. Environ Sci Pollut Res 23(10):9871–9890. https://doi.org/10.1007/s11356-016-6189-0
7. Zaaboub N, Helali MA, Martins MVA et al (2016) Assessing pollution in a Mediterranean lagoon using acid volatile sulfides and estimations of simultaneously extracted metals. Environ Sci Pollut Res 23:21908–21919. https://doi.org/10.1007/s11356-016-7431-5
8. Gao X, Song J, Li X et al (2020) Sediment quality of the Bohai Sea and the northern Yellow Sea indicated by the results of acid-volatile sulfide and simultaneously extracted metals determinations. Mar Pollut Bull 155(3):111147. https://doi.org/10.1016/j.marpolbul.2020.111147
9. Zhuang W, Gao X (2013) Acid-volatile sulfide and simultaneously extracted metals in surface sediments of the southwestern coastal Laizhou Bay, Bohai Sea: concentrations, spatial distributions and the indication of heavy metal pollution status. Mar Pollut Bull 76(1–2):128–138. https://doi.org/10.1016/j.marpolbul.2013.09.016
10. Prica M, Dalmacija B, Rončević S, Krčmar D, Bečelić M (2008) A comparison of sediment quality results with acid volatile sulfide (AVS) and simultaneously extracted metals (SEM) ratio in Vojvodina (Serbia) sediments. Sci Total Environ 389(2–3):235–244. https://doi.org/10.1016/j.scitotenv.2007.09.006
11. Chai M, Shen X, Li R, Qiu G (2015) The risk assessment of heavy metals in Futian mangrove forest sediment in Shenzhen Bay (South China) based on SEM-AVS analysis. Mar Pollut Bull 15, 97(1–2):431–439. https://doi.org/10.1016/j.marpolbul.2015.05.057
12. Li F, Lin J, Liang Y, Gan H, Zeng X, Duan Z, Liang K, Liu X, Huo ZH, Wu C (2014) Coastal surface sediment quality assessment in Leizhou Peninsula (South China Sea) based on SEM–AVS analysis. Mar Pollut Bull 84(1–2):424–436. https://doi.org/10.1016/j.marpolbul.2014.04
13. Gao SP, Hu KD, Hu LY, Li YH, Han Y, Wang HL et al (2013) Hydrogen sulfide delays postharvest senescence and plays an antioxidative role in fresh-cut kiwifruit. Hort Science 48:1385–1392. https://doi.org/10.21273/HORTSCI.48.11.1385
14. Jingchun L, Chongling Y, Spencer KL, Ruifeng Z, Haoliang L (2010) The distribution of acid-volatile sulfide and simultaneously extracted metals in sediments from a mangrove forest and adjacent mudflat in Zhangjiang Estuary, China. Mar Pollut Bull 60:1209–1216. https://doi.org/10.1016/j.marpolbul.2010.03.029
15. Folk RL, Ward WC (1957) Brazos River bar: a study in the significance of grain size Parameters. J Sediment Petrol 27:3–26. https://doi.org/10.1306/74D70646-2B21-11D7-8648000102C1865D
16. Molnia BF (1974) A rapid and accurate method for the analysis of calcium carbonate in small samples. J Sediment Petrol 44:589–590
17. Loring DH, Rantala RTT (1992) Manual for geochemical analysis of marine sediments and suspended particulate matter. Earth Sci Rev 32:235–283. https://doi.org/10.1016/0012-8252(92)90001-A
18. Carritt DE, Carpenter JH (1966) Comparison and evaluation of currently employed modifications of Winkler method for determining dissolved oxygen in seawater. ANASCO report. J Mar Res 24(3):286–318
19. Fujimori K, Ma WL, Kawakami TM, Shibutani Y, Takenata N, Bankow H, Maeda Y (2001) Chemiluminescence method with potassium permanganate for the determination of organic pollutants in seawater. Anal Sci 17:975–978. https://doi.org/10.2116/analsci.17.975
20. Ajayi SO, Vanloon GW (1989) Studies on redistribution during the analytical fractionation of metals in sediments. Sci Total Environ 87–88:171–187. https://doi.org/10.1016/0048-9697(89)90233-7

21. Allen HE, Fu G, Deng B (1993) Analysis of acid-volatile sulfide (AVS) and simultaneously extracted metals (SEM) for the estimation of potential toxicity in aquatic sediments. Environ Toxicol Chem 12:1441–1453. https://doi.org/10.1002/etc.5620120812
22. Yin H, Deng J, Shao S et al (2011) Distribution characteristics and toxicity assessment of heavy metals in the sediments of Lake Chaohu, China. Environ Monit Assess 179:431–442. https://doi.org/10.1007/s10661-010-1746-3
23. Hsieh YP, Chung S-W, Tsau Y-J, Sue C-T (2002) Analysis of sulfides in the presence of ferric minerals by diffusion methods. Chem Geol 182(2–4):195–201. https://doi.org/10.1016/s0009-2541(01)00282-0
24. Chen Y, Shen L, Huang T, Chu Z, Xie Z (2020) Transformation of sulfur species in lake sediments at Ardley Island and Fildes Peninsula, King George Island, Antarctic Peninsula. Sci Total Environ 10(703):135591. https://doi.org/10.1016/j.scitotenv.2019.135591
25. Charriau A, Lesven L, Gao Y, Leermakers M, Baeyens W, Ouddane B et al (2010) Trace metal behaviour in riverine sediments: role of organic matter and sulfides. Appl Geochem 26:80–90. https://doi.org/10.1016/j.apgeochem.2010.11.005f
26. Ngiam LS, Lim PE (2001) Speciation patterns of heavy metals in tropical estuarine anoxic and oxidized sediments by different sequential extraction schemes. Sci Total Environ 275:53–61. https://doi.org/10.1016/s0048-9697(00)00853-6
27. ANZECC & ARMCANZ (2000) Australian guidelines for water quality monitoring and reporting. National water quality management strategy paper no 7, Australian and New Zealand Environment and Conservation Council & Agriculture and Resource Management Council of Australia and New Zealand, Canberra
28. Gammons CH, Drury WJ, Li Y (2000) Seasonal influences on heavy metal attenuation in an anaerobic treatment wetlands facility, Butte, Montana. In: Proceedings from the fifth international conference on acid rock drainage, Society for Mining, Metallurgy, and Exploration, Inc. (SME), Denver, CO, May 20–26, 1159–1168.
29. Tessier A, Campbell P.G.C, Bisson M (1979) Sequential extraction procedure for the speciation of particulate heavy metals. Anal. Chem 51(7):844–851.
30. Kersten M, Forstner U (1987) Effect of sample pretreatment on the reliability of solid speciation data of heavy metals: implications for the study of early digenetic processes. Mar. Chem 22:299–312.

Chapter 6
Update, Conclusions, Recommendations, Future Perspective and Challenges

6.1 Update

The modified risk quotient and Microtox® toxicity according [1], must take into consideration AVS level in addition to geochemical parameters. So, stepwise multiple linear regression (MLR) models should be tested over wide ranges of AVS, OC, grain size and heavy metals as toxicity prediction approach. Although acid volatile sulfide has been identified as a key regulator of metal bioavailability in anoxic sediments, it is unclear how this compound affects metal accumulation in the environment. So, to support the AVS theory an assessment of the connection between AVS (also including SEM, the sulfide-associated metals) and trace metal accumulation by benthic organisms are highly needed and many researchers tried to highlighted this topic [2]. Many processes included biotic and abiotic could remobilize metals from sediments as a second pollution source which cause bioaccumulation and transfer exchanges in the trophic chain represented another challenge that update the vision to metal-sulfide theory [2, 3]. Multiple linear regression-based approach may facilitate the adoption of updated evaluation of metal bioavailability including acid volatile sulfide phenomenon.

6.2 Conclusions

The sedimentary sulfide framework is so dynamic in aquatic environments. AVS is generally one of the most significant and reactive phases. The bioavailability and toxicity of metals in sediments were formerly determined solely by sediment metal concentrations. However, taking into account the impact of geochemical factors (AVS) on metal bioavailability enhanced toxicity prediction. The AVS–SEM models have been proven to be useful in assessing ecological risk. In these models, the relationship between organic matter, AVS, and SEM were used to figure out which fraction was responsible for the metals' toxicity. Metal-associated risks vary even

G. M. El Zokm, *Ecological Quality Status of Marine Environment*, Earth and Environmental Sciences Library, https://doi.org/10.1007/978-3-031-29203-3_6

at constant metal or SEM concentrations due to geochemical heterogeneity in environmental variables such as redox potential, pH, and organic matter. In terms of regional variability, trace metal availability differs from total or normalized trace metal concentrations. $\sum$SEM/AVS, $\sum$SEM-AVS, $\sum$SEM-AVS/f_{OC} models are the most qualified for evaluating hazardous effects on aquatic environments. Sampling, storage, and handling play dramatic roles in changing the AVS signature.

The biotic ligand theory, equilibrium partitioning theory, and the hypothesis of sulfide-bound metals were established and understood to account for metal bioavailability assessed by AVS–SEM models. Integrated approaches are preferred to assess the quality of the aquatic environment; AVS–SEM models, sequential fractionation technique, and sediment toxicity guidelines [4, 5]. By applying the water quality criterion (WQC), the fractionation approach (metal residual fraction; $[M]_R$) and the AVS basis (metal sulfide; $[M]_{AVS}$), the sediment toxicity could be evaluated using the following equation: SQC = KPWQC + [M]R + [M]AVS [6]. Statistical analysis plays a vital role in the prediction of AVS models. Finally, rather than measuring sediment toxicity based on metal concentrations as a total in sediment, SEM in the sediment that is concurrently released by the AVS extraction is a useful approach for assessing divalent cationic metals (Zn, Cu, Cd, Ni, Pb, and CO) pollution. As a result, AVS has been identified as a critical partitioning step in regulating cationic metal bioavailability in sediments.

According to [1], using stepwise multiple linear regression (MLR) models that have been tested over wide ranges of particle size, AVS and TOC, it easier to adopt updated site-specific metals standards that more properly take into account the factors impacting metal bioavailability than using a metal concentration standard alone.

6.3 Recommendations

1. More study is needed to identify the functional link and mechanism of action between heavy metal toxicity in sediments and bioaccumulation, as well as the numerous factors that control this relationship, including metal-binding phases in sediments.
2. Because AVS dissolved species are complex, changeable, and extremely sensitive to environmental variables, particularly oxidation/reduction potential, a deep insight into the geochemistry of the marine sediment must be considered during analysis.
3. Seasonal changes in AVS-SEM systems and toxicity to aquatic biota require a conceptual framework.
4. The integration of SQGs and AVS-SEM models revealed that a single method for toxicity evaluation may be insufficient within the idea of defining quality standards. However, several evaluation methods were compared; the findings were inconsistent, showing that each approach has its own limits.
5. In the next investigation, biological effect evaluations (bioassay) will be necessary for addition to chemical analyses.

6. As a result of sediment resuspension, it is highly suggested that AVS-SEM models are periodically applied to follow the metals, behavior in aquatic environments.
7. Multiple-year studies in lakes could be designed to assess trends in SEM-AVS models with an appropriate spatial scale.
8. Effective implementation strategies are strongly advised to reduce any potential health risks associated with this heavy-metal-based pollution.

6.4 Future Perspective

This research showed that additional investigation is necessary to deepen our understanding of metal-AVS connections under settings of varied SEM and AVS levels. Future efforts should adopt these strategies that connect together for interpretations of metal behaviour (and potential predictions) made possible by AVS models, rather than research that only focuses on using the AVS models to forecast toxicity when appropriate, in order to better characterize the real role of AVS. In the event that successful water quality restoration activities take place, these approaches might enhance interpretations of what can be expected for both future oligotrophication and eutrophication situations.

6.5 Challenges

Although the AVS–SEM approach has been shown to be an effective tool for assessing ecological risk, it has certain limitations, problems, and deficiencies.

1. One of the most significant problems with the AVS–SEM assessment approach is that AVS concentrations might fluctuate as environmental pressures (pH, Eh, oxygen content, etc.) change, leading to increased biological toxicity in sediments that previously had no biological toxicity. Some of the sulfides detected as AVS appear to have been oxidized during elutriation, potentially liberating metals into the water column. As a result, this theory would be acceptable for metal remobilization research only if sulfides are not oxidized during elutriation.
2. Many studies reported that benthos may accumulate metals even so when [SEM] – [AVS] < 0 because their major food source is sediment particles, independent of AVS. Because this conclusion contradicts the categorization criteria established by the USEPA in 2004, this model requires more investigation.
3. Although it is desirable that AVS exceeded SEM in SEM-AVS models, AVS at too high concentration is equally harmful to aquatic species. AVS is commonly found in organic-rich and reductive settings since it is predominantly generated by sulfate reduction in the metabolism of SRB, which are anaerobic bacteria. Benthic creatures and fish may die as a result of the severe anoxic environment.
4. Many studies have found that even when the concentration of SEM in sediments was significantly higher than that of AVS, it did not cause toxicity

in aquatic biota. Other metal-binding phases in sediments, such as organic compounds Mn/FeOx and were shown to reduce bioavailability significantly. However, a comprehensive picture of the sediment's geochemical composition is recommended.

5. There is currently a lack of knowledge of the nature and significance of AVS in the marine sector. The difficulty is around the specification of AVS components in light of the fact that sulfides are extremely vulnerable to oxidation. To resolve the complexity of natural systems, the technique of extracting AVS from sediment requires a sequential leaching strategy. Because diffusion, purge, and trap are mostly ex-situ methods, the outcomes are susceptible to differences in handling procedures.
6. Various settings necessitate different approaches, and shipboard sampling may differ from onshore sampling.

References

1. Ju Y-R, Chen C-F, Lim Y, Tsai C-Y, Chen Ch W, Dong Ch D (2022) Developing ecological risk assessment of metals released from sediment based on sediment quality guidelines linking with the properties: a case study for kaohsiung Harbor. Sci Total Environ 852:158407. https://doi.org/10.1016/j.scitotenv.2022.158407
2. Li F, Lin J, Liang Y, Gan H, Zeng X, Duan Z, Wu C (2014) Coastal surface sediment quality assessment in Leizhou Peninsula (South China Sea) based on SEM–AVS analysis. Mar Pollut Bull 84(1–2):424–436. https://doi.org/10.1016/j.marpolbul.2014.04
3. Brito GB, da Silva Júnior JB, Dias LC, de Santana SA, Hadlich GM, Ferreira SLC (2020) Evaluation of the bioavailability of potentially toxic metals in surface sediments collected from a tropical river near an urban area". Mar Pollut Bull 156:111215. https://doi.org/10.1016/j.marpolbul.2020.111215
4. El Zokm GM, Okbah MA, Younis AM (2015) Assessment of heavy metals pollution using AVS-SEM and fractionation techniques in Edku Lagoon sediments, Mediterranean Sea. J Environ Sci Health Part A Toxic/Hazard Subst Environ Eng 50:571–584. https://doi.org/10.1080/10934529.2015.994945
5. Okbah MA, Younis AM, El Zokm GM (2015) Heavy metals fractionation and acid volatile sulfide (AVS) in the Bardawil lagoon sediments, northern Sinai, Egypt. Dev Analyt Chem 2. https://doi.org/10.14355/dac.2015.02.001
6. USEPA (1994) Methods for measuring the toxicity and bioaccumulation of sediment-associated contaminants with freshwater invertebrates. 600/R-94/024. Washington, DC

Bibliography

7. Brito GB, da Silva JJB, Dias LC, de Santana SA, Hadlich GM, Ferreira SLC (2020) Evaluation of the bioavailability of potentially toxic metals in surface sediments collected from a tropical river near an urban area. Mar Pollut Bull 156:111215. https://doi.org/10.1016/j.marpolbul.2020.111215
8. Chapman PM, Wang F, Adams WJ, Green A (1999) Appropriate applications of sediment quality values for metals and metalloids. Environ Sci Technol 33:3937–3941. https://doi.org/10.1021/es990083n

9. Fu L, Hu J, Shen W, Huang X, Luo J, Jia M, Zhang J (2014) Occurrence and implications of SEM-AVS for surface sediments from Baihua Lake, China. Soil Sediment Contamination Int J 23(3):287–312. https://doi.org/10.1080/15320383.2014.826621
10. El Zokm GH, Masoud MS, El-Shorbagi EK, Elsamra RMI, Okbah MA (2023) Reactive sulfide dynamic models for predicting metal hazardous in sediments of two northern Egyptian lakes. Mar Pollut Bull 188:114694. https://doi.org/10.1016/j.marpolbul.2023.114694
11. Hansen DJ, Berry WJ, Boothman WS, Pesch CE, Mahony JD, Di Toro DM, Yan Q (1996) Predicting the toxicity of metal-contaminated field sediments using interstitial concentration of metals and acid-volatile sulfide normalizations. Environ Toxicol Chem 15(12):2080–2094. https://doi.org/10.1002/etc.5620151204
12. Howard DE, Evans RD (1993) Acid-volatile sulfide (AVS) in a seasonally anoxic mesotrophic lake: Seasonal and spatial changes in sediment AVS. Environ Toxicol Chem 12:1051–1057. https://doi.org/10.1002/etc.5620120611
13. Lasorsa B, Casas A (1996) A comparison of sample handling and analytical methods for determination of acid volatile sulfides in sediment. Mar Chem 52:211–220. https://doi.org/10.1016/0304-4203(95)00074-7
14. Lee JS, Lee JH (2005) Influence of acid volatile sulfides and simultaneously extracted metals on the bioavailability and toxicity of a mixture of sediment-associated Cd Ni, and Zn to polychaetes Neanthes arenaceodentata. Sci Total Environ 338(3):229–241. https://doi.org/10.1016/j.scitotenv.2004.06.023
15. Li D, Liu X, Liu Z et al (2016) Variations in total organic carbon and acid-volatile sulfide distribution in surface sediments from Luan River Estuary, China China. Environ Earth Sci 75:1073. https://doi.org/10.1007/s12665-016-5873-1
16. Liu J, Pellerin A, Antler G, Kasten S, FindlayAJ DI, JørgenseB B (2020) Early diagenesis of iron and sulfur in Bornholm Basin sediments: the role of near-surface pyrite formation. Geochim Cosmochim Acta. https://doi.org/10.1016/j.gca.2020.06.003
17. Otero XL, Macias F (2002) Variation with depth and season in metal sulfides in salt marsh soils. Biogeochemistry 61(3):247–268. https://doi.org/10.1023/a:1020230213864
18. Rickard D (1997) Kinetics of pyrite formation by the H2S oxidation of iron (II) monosulfide in aqueous solutions between 25 and 125 °C: the rate equation. Geochim Cosmochim Acta 61:115–134. https://doi.org/10.1016/S0016-7037(96)00321-3
19. Simpson SL, Ward D, Strom D, Jolley DF (2012) Oxidation of acid-volatile sulfide in surface sediments increases the release and toxicity of copper to the benthic amphipod Melita plumulosa. Chemosphere 88:953–961. https://doi.org/10.1016/j.chemosphere.2012.03.026
20. Wu Z, Ren D, Zhou H, Gao H, Li J (2016) Sulfate reduction and formation of iron sulfide minerals in nearshore sediments from Qi'ao Island, Pearl River Estuary, Southern China. Q Int:1–11. https://doi.org/10.1016/j.quaint.2016.06.003
21. Zhang Y, Li H, Yin J, Zhu L (2021) Risk assessment for sediment associated heavy metals using sediment quality guidelines modified by sediment properties. Environ Pollut 275:115844. https://doi.org/10.1016/j.envpol.2020.115844

SPRINGER NATURE

GPSR Compliance

The European Union's (EU) General Product Safety Regulation (GPSR) is a set of rules that requires consumer products to be safe and our obligations to ensure this.

If you have any concerns about our products, you can contact us on ProductSafety@springernature.com

In case Publisher is established outside the EU, the EU authorized representative is:

Springer Nature Customer Service Center GmbH
Europaplatz 3
69115 Heidelberg, Germany

The manufacturer's authorised representative in the EU is Springer Nature Customer Service Centre GmbH, Europaplatz 3, 69115 Heidelberg, Germany. If you have any concerns regarding our products, please contact ProductSafety@springernature.com

Printed and bound by CPI Group (UK) Ltd, Croydon, CR0 4YY
01/10/2026
02225880-0011